Forest and Plant Bioproducts Division

Held at the 2023 AIChE Annual Meeting

Orlando, Florida, USA
5-10 November 2023

ISBN: 978-1-7138-9295-3

Printed from e-media with permission by:

Curran Associates, Inc.
57 Morehouse Lane
Red Hook, NY 12571

Some format issues inherent in the e-media version may also appear in this print version.

Printed with permission by Curran Associates, Inc. (2024)

For permission requests, please contact AIChE
at the address below.

AIChE
120 Wall Street, FL 23
New York, NY 10005-4020

Phone: (800) 242-4363
Fax: (203) 775-5177

www.aiche.org

Additional copies of this publication are available from:

Curran Associates, Inc.
57 Morehouse Lane
Red Hook, NY 12571 USA
Phone: 845-758-0400
Fax: 845-758-2633
Email: curran@proceedings.com
Web: www.proceedings.com

TABLE OF CONTENTS

ADVANCES IN LIGNOCELLULOSE PROCESSING, CONVERSION, AND VALORIZATION I

ADVANCED SEPARATIONS PROCESSES IN BIOPROCESSING AND BIOMATERIALS

ADVANCES IN LIGNIN CHEMISTRY, ISOLATION, AND DEPOLYMERIZATION

ANDREW CHASE AWARD IN HONOR OF DR. DEBORAH MIELEWSKI (INVITED TALKS)

COMPUTATIONAL METHOD AIDED BIOMASS AND WASTE UTILIZATION

BIOMASS CHARACTERIZATION, PRETREATMENT, AND FRACTIONATION I

MODIFICATION, FUNCTIONALIZATION AND PROCESSING OF BIOMATERIALS

DIVISION PLENARY: VALORIZATION OF WASTE PLASTICS (INVITED TALKS)

LIGNIN FOR SUSTAINABLE INDUSTRIAL USES

POSTER SESSION: ADVANCES IN FOREST AND PLANT BIOMASS UTILIZATION

ADVANCES IN LIGNOCELLULOSE PROCESSING, CONVERSION, AND VALORIZATION II

BIODEGRADABLE POLYMERS FROM RENEWABLE AND WASTE RESOURCES AND BIOCOMPOSITES FROM INDUSTRIAL COPRODUCES AND NATURAL FIBERS

CATALYTIC AND THERMOCHEMICAL CONVERSION OF LIGNOCELLULOSIC MATERIALS I

HIGH-VALUE AND PLATFORM CHEMICALS FROM RENEWABLE RESOURCES AND WASTES

CELLULOSE NANO MATERIALS - CRYSTALLINE, FIBROUS, GELS, FILMS AND FOAMS FOR BROAD APPLICATIONS IN MEDICAL, ENERGY AND THE ENVIRONMENT.

BIOMASS CHARACTERIZATION, PRETREATMENT, AND FRACTIONATION II

CATALYTIC AND THERMOCHEMICAL CONVERSION OF LIGNOCELLULOSIC MATERIALS II

364a One Pot Production of Levulinic Acid from Lignocellulosic Biomass Using a Catalytic Membrane Reactor

Zhexi Zhu, Ranil Wickramasinghe, Xianghong Qian

Conversion of cellulosic biomass to levulinic acid (LA), a platform chemical for bioplastics and hydrocarbon fuels has become a major focus for biomass conversion. Currently, most of the levulinic acid production processes from lignocellulosic biomass rely on the corrosive homogeneous acid or toxic metal-based catalyst. Here a polymeric solid acid catalyst is designed and synthesized to directly convert cellulosic biomass to levulinic acid in one step at mild conditions without pretreatment with high selectivity and high conversion. Moreover, the catalyst is immobilized on a membrane substrate which enables the immediate separation of the levulinic acid product from the reactor driving the reaction to completion and improving the yield. Here the production of levulinic acid is optimized by varying the reaction conditions including the solvent, solvent mixture, reaction time and temperature for a variety of biomass feedstock. In particular, green solvent mixtures comprising ionic liquid or water with γ-valerolactone (GVL) are found to improve the selectivity and yield for the levulinic acid production. Levulinic acid yields up to 90% have been achieved at mild reaction conditions from different lignocellulosic biomass including nut shells and husks, corn stover, rice husks, and wheat straws. In addition, our solid acid catalyst is reusable, less corrosive, environmentally friendly and is a promising technology for realizing future biorefinery.

364b Cellulose Dissolution and Regeneration in Calcium Bromide Solution

Jiansong Chen, Zhiqiang Pang, Haishun Du, Xuejun Pan

Cellulose is the most abundant natural polymer in the world. As a promising material, it offers advantages including renewability, biodegradability, biocompatibility, barrier properties, attractive appearance, non-toxicity, and low cost. Efficiently converting this renewable resource into value-added products, such as regenerated cellulose materials and cellulose derivatives, requires the critical step of cellulose dissolution. Inorganic molten salt hydrates, such as lithium bromide (LiBr), zinc bromide ($ZnBr_2$), zinc chloride ($ZnCl_2$), and calcium bromide ($CaBr_2$) solutions, have drawn increasing attention as a new category of cellulose solvents. This study provides a systematic investigation into cellulose dissolution by $CaBr_2$ solutions. The cellulose dissolution by $CaBr_2$ is a process of disassembling the strong matrix of cellulose chains assembled via intermolecular hydrogen bonding (H-bonds) and hydrophobic stacking. Parameters influencing the dissolution of microcrystalline cellulose (MCC), including $CaBr_2$ concentration, MCC concentration, temperature, and time, were investigated. The results showed that 2% MCC can be dissolved within 15 min by 58% $CaBr_2$ at 100 °C, without hydrolysis of cellulose to glucose or further conversion to 5-hydroxymethylfurfural. Additionally, this study prepared cellulose aerogels via a process of dissolution and regeneration. The cellulose aerogels had high porosity, low density, and strong matrix properties, making them potentially useful in adsorption, oil/water separation, thermal insulation, and biomedical applications, among other fields. Overall, this study provides a systematic investigation into cellulose dissolution by $CaBr_2$ solutions and has important implications for the development of new processes to produce cellulose-based materials with a broad range of applications.

364c Phenoxyethanol Pretreatment of Lignocellulosic Biomass Under Biphasic and Monophasic Solvent Systems

Xuesong Tan, Yiqi Zhang, Xinshu Zhuang

Pretreatment is the basis to realize the bio-refining process of preparing fuel and chemicals from lignocellulosic biomass, which determines the direction and efficiency of refining. Organosolv pretreatment has become one of the hot spots in pretreatment research because of its high efficiency of component disassembly, high retention of decomposed cellulose components and high purity of recycled lignin components.

Phenoxyethanol is considered a green, non-toxic organic solvent, with a good ability to dissolve lignin. Due to its water immiscible, we constructed a biphasic phenoxyethanol/water system for rice straw pretreatment[1,2]. After pretreatment, the residue is mainly cellulose, meanwhile, five-carbon sugars decomposed from hemicellulose are dissolved in aqueous phase and lignin is distributed in organic phase. The biphasic pretreatment shows outstanding advantages and broad application prospects in the biorefinery of lignocellulosic substrates due to the separation process for products in the solvent phase, and the reutilization of solvents. The phenoxyethanol/water biphasic system could also be performed under alkaline conditions[3]. However, a small amount of water is still dissolved in the organic solvent that makes up biphasic system, and few studies have focused on the pretreatment performance of this monophasic system.

Herein, the effects of water content in phenoxyethanol solution on fractionated components and enhanced enzymatic hydrolysis of poplar pretreatment under acid condition was investigated. With the decrease of water content, the solvent system changed from biphasic to monophasic. Compared to biphasic system, the monophasic system overcame the mass transfer hindrance and showed better delignification and hemicellulose removal during the pretreatments. Impressively, a one-step pretreatment of poplar below 130°C in monophasic system with lower water content resulted in almost complete removal of lignin and hemicellulose, yielding a residue with ~90% cellulose content, and with ~90% enzymatic digestion. Overall, water was involved not only in changing the phase state of the system, but also in changing the lignin solubilization properties of the solution, determining the strength of the reaction and supporting the hydrolysis of the components. Our findings are useful for organosolv pretreatment and bio-refinings.

References:

1. Screening Solvents Based on Hansen Solubility Parameter Theory To Depolymerize Lignocellulosic Biomass Efficiently under Low Temperature. *ACS Sustainable Chem. Eng.* 2019, 7, 8678.

2. Biphasic fractionation of rice straw under mild condition in acidified 2-phenoxyethanol/water system. *Industrial Crops & Products*. 2020, 145, 112091.

3. A Novel Recyclable Alkaline Biphasic 2-Phenoxyethanol/Water System for Rice Straw Biorefinery under Mild Conditions. *ACS Sustainable Chem. Eng.* 2020, 8, 7649.

4. A one-step deconstruction-separation organosolv fractionation of lignocellulosic biomass using acetone/phenoxyethanol/water ternary solvent system. Bioresource Technology. 2021, 342, 125963.

5. Mild organosolv pretreatment of sugarcane bagasse with acetone/phenoxyethanol/water for enhanced sugar production. Green Chemistry. 2023, 25, 1169.

364d Cellulose Hydrolysis Catalyzed By Solid Acid to Levulinic Acid

Suping Zhang

As a clean renewable energy, biomass energy is of great significance to environmental protection, especially "carbon reduction". In this thesis, a highly efficient catalytic system for producing levulinic acid was developed to promote the conversion of cellulose. Lignosulfonate was activated by phosphoric acid, carbonized and sulfonated by sulfanilic acid to prepare lignin-based solid acid catalyst with high specific surface area and high -SO_3H load. It was proved that more phosphorus-containing groups would be grafted on the catalyst surface at high impregnation ratio and high temperature, which provided more grafting sites for -SO_3H. In the biphasic reaction system (V_{MIBK}:V_{H2O}=1:1, 0.2g/mL NaCl), the highest yield of levulinic acid was 67.93% under the conditions of reaction temperature of 190°C, reaction time of 150min, PLSA4-700 loading of 50wt% and substrate loading of 2wt%. In addition, the catalyst showed excellent catalytic performance and good cycling.

364e Extraction and Characterization of Cellulose Nanocrystals from Corn Husk Wastes and It´s Application in Bioplastic Preparation.

Sergio Mayta Paucara, Maria Quintana

Maize is the most cultivated cereal in the world. In fact, according to data of the Food and Agricultural Organization of the United Nations, global production reached up to 1.16 billion tons in 2020. In Peru, according to data from the Minister of Agriculture (MIDAGRI, 2021), more than 414 thousand tons of corn were harvested in 2020, mainly in the coastal valleys of Lima and Ica regions.

In this context, cereals production has also given rise to considerable amounts of waste, which are the residues generated by the harvesting and processing of the feedstock. In particular maize wastes include stalks, chaffs, steams, silks, and husks; out of these components' stalks constitute more than half of the stover biomass while the husk accounts for 10% of the total mass. Furthermore, these wastes barely degrade on the soil surface, making their disposal still a challenge, especially in developing countries where the lack of efficient and environmentally friendly techniques ultimately generates environmental threats.

In this sense, open air burning is one of the main disposal strategies for corn waste in Peru, however, this practice has been widely questioned by the scientific community as it constitutes one major source of gaseous and particulate emissions to the atmosphere. In fact, both greenhouse gasses (carbon dioxide (CO2), carbon monoxide (CO), ie.) and volatile organic compounds (ammonia (NH3), sulfur dioxide (SO2), ie.) would get released to the surroundings.

On the other hand, scientists have shown a particular interest in the valorization of different lignocellulosic biomass over the last decade, as it constitutes the largest source of renewable organic material on Earth. In particular, cellulose has received remarkable attention considering its outstanding mechanical properties (tensile strength between 5 and 7 GPa along the chain direction) and its versatility in the synthesis of added-value derivates.

Moreover, micro and nano forms of cellulose are quite relevant because of their unique properties conferred at such a small scale. Cellulose nanocrystals (NCC) are rod-like particles with typical widths and lengths in the range of 5–70 nm and several micrometers, respectively. NCC exhibits low density (~1.46 g/cm3), an outstanding mechanical performance (bending strength estimated at 10 GPa and elastic modulus ranging from 10 and 100 to 200 Pa, respectively), and a high surface area/aspect ratio of 10 up to 70.

Under this context, this research focuses on the valorization of Peruvian corn wastes into cellulose nanocrystals using both mechanical and chemical treatments combined.

The feedstock was collected from local markets in Lima, Peru, and after washing it several times, the sample was dried, milled, and sieved under 850 μm. Afterward, dried corn husk powder was treated with 4% (w/v) sodium hydroxide twice. Then, peracetic acid bleaching was conducted according to a previous methodology. This cellulose sample was characterized by instrumental techniques and gravimetric measurements. Finally, acid hydrolysis was conducted with concentrated sulfuric acid at 45°C followed by centrifugation, dialysis and low-frequency ultrasonication. The nanocrystalline cellulose suspension obtained was used to prepare films by simple casting method.

The cellulose was extracted with a 79.4% yield while its purity was estimated at 70.3%. On the other hand, in order to identify characteristic functional groups of spectral bands, characterization by FT-IR spectroscopy was carried out, and characteristic bands of cellulose were found at 1426 cm-1, 1325 cm-1, 1163 cm-1, and 1110 cm-1. Likewise, the decrease in signals at 1251 cm-1 and 1736 cm-1 indicates that hemicellulose and lignin were removed after the chemical treatments.

At the same time, x-ray diffractograms showed peaks at 22.5° and 34°, characteristics of the crystal polymorphs of cellulose, while the crystallinity index was estimated in 54 by the Segal equation. SEM images showed a change on the structure of the material from irregular and rough to a highly fibrillated network arrangement.

The average hydrodynamic diameter of the isolated CNC was determined by dynamic light scattering (DLS). The nanocrystals dimension was found to be 268 nm, however, its important to highlight that the hydrodynamic diameter represents the dimension of an equivalent sphere for different types of particles (individual size, aggregates, and agglomerates) and thus may not represent the actual size of the particle, in fact, CNC is known for having a high length: width ratio (aspect ratio). Therefore, aiming to obtain a better understanding of the morphology of the crystals, atomic force microscopy was performed into diluted suspensions, it was found that the nanocrystals had a rod-like shape, with an average aspect ratio of 35.

Furthermore, for comparison, commercial nano cellulose was also prepared and the dimension was found to be 218 nm.

It is also important to mention that an interesting effect caused by low-frequency ultrasonication was noticed. As the sonication time increased both the viscosity and the particle size of the nanocrystalline suspension decreased. This effect might be a consequence of the rearranging of the particles caused by cavitation. Previous studies have found that ultrasound induces a decrease in the aspect ratio of suspension while promoting the individualization of particles. Thus, further research studying these effects is still needed.

CNCs films were cast by pouring the suspension into a petri dish and drying overnight. These films were highly transparent and brittle. Their resistance to both alkaline and acidic medium was evaluated by measuring the weight loss over the days and comparing it to polyethylene films.

Regarding the water absorption capacity, the films were able to absorb their weight up to 3 times, approximately. It was noted that the bioplastics obtained had good resistance to acids (sulfuric acid) and bases (sodium hydroxide) up to concentrations of 25% and 40%, respectively. After 5 days there was no loss in the mass of the CNCs films. When these films are compared with those commercially obtained (polyethylene), it was observed that the latter lose 20% of their mass in the presence of 40% alkali, while the former was intact.

This evidence could respond to the high stability of the crystalline structure and the strong bonding between molecules, making it suitable for future applications in the medical and food industries, where these properties are highly appreciated.

364f Carbon Dioxide and Organic Base Additives for Polar Aprotic Organosolv Pulping of Lignocellulosic Biomass

Kelechi A. Agwu, Rae Belmont, Jayna M. Enguita, James D. Sheehan

Lignocellulose is the most abundant terrestrial biomass on earth, with over 181.5 billion tons produced annually.[1] Lignocellulosic biomass is comprised of three major macromolecular components, which include cellulose (40-60 wt%), hemicellulose (10-40 wt%), and lignin (15-30 wt%).[2] Collectively the constituent macromolecules form a lignin-carbohydrate complex that facilitates the structural integrity and chemical recalcitrant of plant tissues. The chemical recalcitrance of lignocellulose can be overcome by thermochemical pretreatments which fractionate lignocellulose into its constituent components. Chemical pretreatments cleave lignin-carbohydrate linkages, solubilize and fractionate lignins, and produce cellulose-rich pulps. Historically, chemical pulping of lignocellulose (e.g., kraft pulping) has focused on solely isolating the cellulose fraction for producing paper pulp. Lignin fractions are removed as black liquor, a mixture consisting of condensed lignins and pulping chemicals, which is subsequently incinerated in order to generate low-grade heat and recoup the pulping reagents.[2] Alternatively, as lignins are the largest biogenic source for aromatic compounds, "lignin-first" pulping methods, such as organosolv pulping, focus on the isolation of lignin fractions in order to enhance the economic viability of lignocellulose valorization processes.[3] However, organosolv pulping finds limited commercialization due to the volatility of the organic solvents (e.g., methanol, ethanol, acetone) which require elevated pressures (15-35 bar) and specialized equipment to keep the solvent species in a condensed liquid phase.[4,5]

In the present study, polar aprotic solvents and organic bases with low vapor pressures are evaluated for the organosolv pulping of lignocellulosic biomass. Carbon dioxide (CO_2) atmospheres are also evaluated as additives to facilitate delignification. Under mild thermal and pressure conditions ($T < 210$ °C, < 10 bar), high degrees of delignification (80-90 wt%) of softwood species are observed with 2 hours. The influence of process conditions (e.g., temperature, pressure, time) and solvent composition on the organosolv pulping process is evaluated through characterization of the isolated cellulose and lignin fractions via spectroscopic, microscopic, and wet chemistry analytical methods. The outcomes of the present research advance the development of mild and scalable chemical pretreatments for fractionating lignocellulose into its constituent polymers, which can serve as feedstocks for value-added, biobased materials.

References

(1) Dahmen, N.; Lewandowski, I.; Zibek, S.; Weidtmann, A. Integrated Lignocellulosic Value Chains in a Growing Bioeconomy: Status Quo and Perspectives. *Glob. Chang. Biol. Bioenergy* **2019**, *11*, 107–117. https://doi.org/10.1111/gcbb.12586.

(2) Schutyser, W.; Renders, T.; Van den Bosch, S.; Koelewijn, S. F.; Beckham, G. T.; Sels, B. F. Chemicals from Lignin: An Interplay of Lignocellulose Fractionation, Depolymerisation, and Upgrading. *Chem. Soc. Rev.* **2018**, *47*, 852–908. https://doi.org/10.1039/c7cs00566k.

(3) Abu-Omar, M. M.; Barta, K.; Beckham, G. T.; Luterbacher, J. S.; Ralph, J.; Rinaldi, R.; Roman-Leshkov, Y.; Samec, J. S.; Sels, B. F.; Wang, F. Guidelines for Performing Lignin-First Biorefining. *Energy Environ. Sci.* **2021**, *14*, 262–292. https://doi.org/10.1039/d0ee02870c.

(4) Galbe, M.; Wallberg, O. Pretreatment for Biorefineries: A Review of Common Methods for Efficient Utilisation of Lignocellulosic Materials. *Biotechnol. Biofuels* **2019**, *12* (1), 1–26. https://doi.org/10.1186/s13068-019-1634-1.

(5) Brosse, N.; Hussin, M. H.; Rahim, A. A. Organosolv Processes. *Adv. Biochem. Eng. Biotechnol.* **2019**, *166* (March 2017), 153–176. https://doi.org/10.1007/10_2016_61.

364g Study of Physical Properties and Multi-Phase Equilibrium of Ternary Mixture (Brine, Oil, Green Corrosion Inhibitor from Lignocellulosic Resource)

Tianxing Cai

Lignocellulosic resource is the biomass material with polysaccharides and aromatic polymers. This bio-source based chemical and material can be capable of producing value ended products such as sustainable corrosion inhibitors. Green corrosion inhibitors are biodegradable and do not contain heavy metals or other toxic compounds. The physical properties such as temperature, concentration, solubility and multi-phase equilibrium has a great impact on the inhibition effect. The experiment measurement and modeling techniques are used to study the effect of multiple physical properties and multi-phase equilibrium of ternary mixture (Green Corrosion Inhibitor, Brine, Oil). Since the corrosion is still one of the industrial challenges to the equipment, pipeline and other infrastructure, the results from this study will benefit the environment, chemical industry, and give contribution to knowledge in the field toward the anti-corrosion efficiency of carbon steel for the transportation of brine and crude oil.

364h Enhanced Carbon Dioxide Capture By Functionalizing Activated Hydrochar Derived from Loblolly Pine with Deep Eutectic Solvent

Swarna Saha, Sarah Pezzenti, Toufiq Reza

CO_2 emission has been a burning question in recent years which results in global warming and other natural calamities. Loblolly pine (LP), being a common forest remnant could be transformed into cost effective adsorbent to capture CO_2 by functionalizing with deep eutectic solvent (DES) known as a potential green solvent. This research aims to produce ultraporous activated hydrochar obtained from loblolly pine by hydrothermally carbonizing at temperatures ranging from 200°C to 260°C after 1 hour of pyrolysis at 600°C which is further functionalized with choline chloride-urea, choline chloride-ethylene glycol, tetra-n-butyl ammonium bromide-urea and contrast the efficiency of CO_2 capture with activated hydrochar from LP without post treatment. The activated chars exhibited high surface porosity (118.34-356.78 m^2/g) and pore volume (0.074-0.1573 cm^3/g), and after functionalization with DES, the adsorbent demonstrated a boost in nitrogen percentage and favorable surface functionalities. While comparing the CO_2 capture capacity at 3 bar and 25°C, functionalized activated chars demonstrated improved uptake efficiency. The combined effect of higher surface functionalities and high porosity reflected a rise in CO_2 capture effectiveness through improving both physical and chemical adsorption mechanism.

251a Fibrillated Cellulose Fibers for Barrier Coating of Fruits to Extend Shelf Life

Jing Geng, Xiao Zhang, Junyong Zhu

This study for the first time evaluated the feasibility of using 100% microfibrillated cellulose (MFC) for barrier coating to preserve the shelf life of bananas. MFC was produced from a commercial bleached kraft hardwood pulp through mechanical milling and barrier coating was accomplished through dipping bananas in aqueous MFC suspensions. The performance of barrier coating on banana peel surface was evaluated from visual observations, banana weight loss, firmness, and dissolved soluble sugars content. The results indicate that coating MFC on banana peel surface delayed the banana browning, softening, and conversion of banana carbohydrates into soluble sugars. Furthermore, the extent of mechanical fibrillation in producing MFC affected MFC suspension wettability of banana peel surface, as well as the effectiveness of banana preservation. Casting-made MFC films were used to understand various factors on MFC barrier coating performance. The effects of the extent of MFC fibrillation was found to be more effective than increasing the amount of coating in reducing water vapor permeation for better fruit protection.

Keywords: Barrier coating; Microfibrillated cellulose; Shelf life; Preservation; Vapor permeation

251b Counter-Current Chromatography for Separating Lignin Monomers from Reductive Catalytic Fractionation Oils from Various Feedstocks

Hoon Choi, Manar Alherech, Hakan Olcay, Louis Chirban, Sean Woodworth, Hannah Alt, Eric Tan, Gregg Beckham

Lignin-first biorefining aims to fractionate lignocellulosic biomass into a polysaccharide-rich pulp and a stabilized lignin oil. Among lignin-first biorefining methods, reductive catalytic fractionation (RCF) has received substantial attention as a process to selectively extract lignin from the plant cell wall, cleave aryl-ether linkages, and catalytically stabilize the resulting reactive aromatic intermediates using a hydrogen donor. RCF has been shown to be a feedstock-agnostic process, able to effectively extract and depolymerize lignin from hardwoods, softwoods, and grasses into oils rich in monomers and oligomers. Converting these products into valuable chemicals and materials is reliant on the development of advanced separation processes capable of monomer-monomer and monomer-oligomer separations with minimal energy/chemical inputs. To that end, here we demonstrate that counter-current chromatography (CCC) is a promising chromatographic technique for this application. In this presentation, we will discuss the separation and isolation of lignin-derived monomers using CCC from RCF oils of softwoods, hardwoods, and grasses, which exhibit key differences in lignin chemistry that dictate solvent choice. Based on the experimental data, a techno-economic analysis of the downstream process using CCC will be presented and compared with other separation approaches.

251c Dynamic Capture and Elution of mRNA with Affinity Membranes for Vaccine Production

Thomas Neuman, Riddhi Banik, Miral Al Sharabati, Zerui Hao, Andrew Hartin, Todd Przybycien, James (Chip) Kilduff, Georges Belfort

Although mRNA-based vaccines are known for their protection against SARS-CoV-2, they have applications well beyond this disease. With efficacies >95%, mRNA vaccines are highly translatable, noninfectious, and unlikely to integrate into a patient's genome. Further increasing the applicability and accessibility of RNA therapies depends on overcoming downstream purification processing limitations. Industry currently relies on diffusion limited chromatography which is inefficient with respect to residence time, reduces final product integrity, requires large footprint and thus, impacts process economics and efficiency. This work describes a convective-based oligo-dT affinity membrane separation as an improved option to reduce the residence time of downstream operations while allowing for continuous processing and high retention of folded mRNA.

As part of a global study investigating affinity membrane fabrication, performance and modeling of mRNA capture, this presentation will focus primarily on experimental quantification and optimization of affinity membrane effectiveness in terms of dynamic capture and elution of mRNA. The development of a High-Performance Liquid Chromatography (HPLC) adaptable method will be described, resulting in membrane binding capacities of ~4.0 mg/mL of Firefly Luciferase (Fluc)-mRNA, with near 100% recovery of bound nucleic acids, and strong maintenance of product integrity in eluent samples. Furthermore, several optimization studies will be reported involving ligand surface concentration, ligand type and length, and feed conditions to maximize the recovery of bound nucleic acids while balancing experimental cost (**Fig. 1**).

Overall, results indicate that the modified RC membranes provide an improved alternative to current industrial separations. The membrane system is able to leverage a convective nature, hence speed of operation, to provide high mRNA capacity offering attractive production economics, while maintaining recovery and purity across operational conditions. Ultimately, further optimization of the affinity system and scale-up to other formats such as hollow fibers will pave the way to increasing the accessibility and application range of exciting, mRNA-based therapeutics.

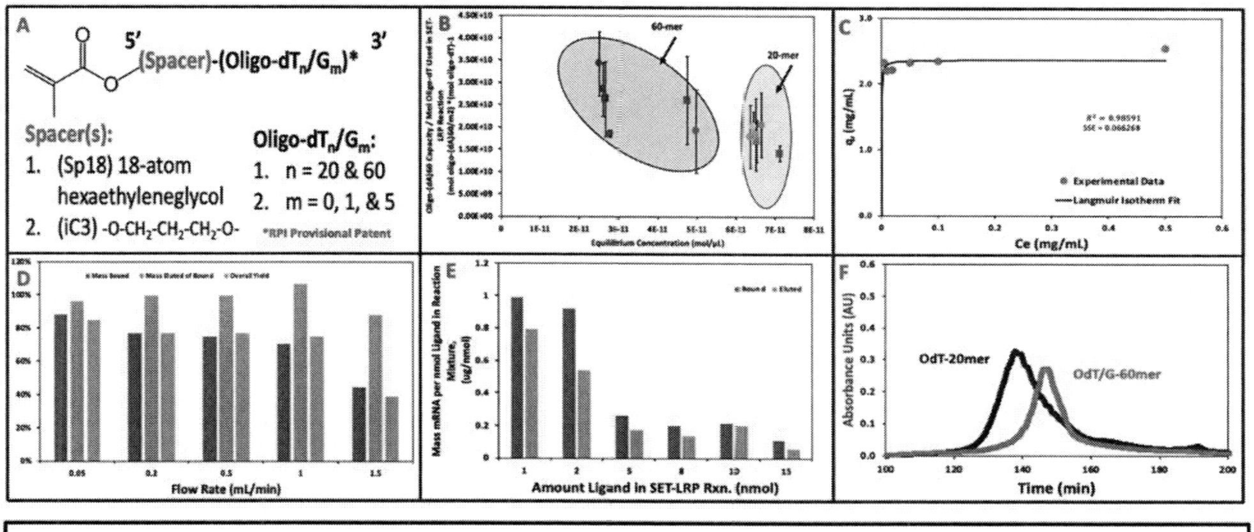

Fig. 1: mRNA Capture Optimization Studies: Influence of membrane/operational conditions on membrane performance for capturing and eluting Fluc-mRNA with a modified microporous flat sheet regenerated cellulose affinity membranes.
(A) Visualization of affinity ligand library of 12 monomer types utilized in performance comparison studies. (B) Static binding capacity results reported for various monomer types described in (A) using a mRNA poly-A tail surrogate Oligo-dA_{60} molecule. The data in (B) led to the choice of two monomers: Oligo-dT_{20} and Oligo-dT/G_{60} for further testing. (C) Capacity (q) in terms of mRNA mass per volume of solid membrane matrix versus feed concentration of mRNA reported for stack of 2 modified membranes, 5 ug Fluc-mRNA injected into system. The data in (C) can be fit with a Langmuir type isotherm. (D) Conservation of percent mass bound, mass eluted of bound and overall yield as a function of operational flow rate (or permeate flux), 10 ug Fluc-mRNA injected into system. (E) Mass of Fluc-mRNA bound and eluted/mole of ligand on the RC membrane surface versus the monomer (Oligo-dT_{20}) amount added to the grafting reaction mixture is shown. The data in (E) shows that the higher the monomer mass in the reaction mixture, the less mass is grafted and eluted per nmol grafted ligand, 5 ug Fluc-mRNA injected into system. (F) Relative HPLC elution profiles reported for stack of 4 modified membranes of Oligo-dT_{20} or Oligo-T/G_{60}, produced by injection of 50 ug Fluc-mRNA. Variation in retention time can be attributed to difference in binding energy with target molecule.

251d Spinning Band Distillation of Pyrolysis Oil Phenolics

Yaseen Elkasabi, Charles A. Mullen, Kerby Jones, Victor Wyatt

Fast pyrolysis of biomass produces organic bio-oils, which can be refined into finished hydrocarbon fuels. Oils of relatively low oxygen content (<15 wt%) can be distilled, leaving behind a solid carbonaceous residue. We've demonstrated how extraction can separate the distillates into a hydrocarbon fraction and a phenolics fraction, the latter consisting predominantly of one-ring phenolics. We've also demonstrated the feasibility of continuous flash distillation for bulk bio-oils and hydrocarbons. For phenolics, more careful distillation methods are warranted to produce desired purities.

In this talk, we discuss the purification and utilization of partially deoxygenated bio-oil fractions using spinning band distillation. We utilized a spinning band column distillation system, designed to perform at 200 theoretical plates. Catalytic (HZSM-5) pyrolysis bio-oils from switchgrass biomass first underwent extraction directly, producing a phenolic fraction and a hydrocarbon fraction. Subsequently, the phenolic fraction underwent a preliminary flash distillation before spinning band distillation, in order to remove residues. Spinning band distillation resulted in fractions with more than 80 wt% phenol (the balance was o-cresol), which can be purified by recrystallization to >93 wt%. We will also discuss the composition of other fractions characterized using NMR. While we've distilled hydrocarbon fractions in the past, none have produced a coke residue fraction. We distilled the hydrocarbon fraction and converted this specific residue into a high-purity coke product. Structural and electrical properties will be presented, including results from XRD, resistivity testing, and XRF.

251e A Scalable, High-Throughput Platform for Isolating Evs, Lipoproteins, and Rnps from Blood Plasma, Urine, and Saliva, Using Isoelectric Fractionation

Himani Sharma, Fnu Vivek, Satyajyoti Senapati, Hsueh-Chia Chang

Extracellular RNA (exRNA) nanocarriers secreted in biological fluids play a crucial role in a variety of complex cellular functions by trafficking signaling molecules from cell to cell. exRNAs abundant in physiological fluid (such as blood, urine, and lymph fluids) are encased and protected by three classes of nanocarriers: Extracellular vesicle (EVs), Lipoprotein (LLPs), and Ribonucleic protein (RNPs). Several recent studies show that exRNA nanocarriers from cancer tumor cells contain a unique set of molecular cargos that can indicate the different stages of tumor progression and hence has the potential to serve as diagnostic and prognostic biomarkers. It is thus very important to separate these carriers and analyze them separately. However, the overlapping size and density of the nanocarriers have so far prevented their efficient physical fractionation, thus impeding independent downstream molecular assays. Moreover, affinity-based and liquid extraction technologies tend to exhibit significant bias and contamination, thus corrupting the quantification of expression level. Hence, there is a need for new technology to provide bias-free, high-throughput, and high-yield nanocarrier fractionation. To address these shortcomings, we take the advantage of their distinct isoelectric points and developed a novel free-flow isoelectric focusing (charge-based) technique to fractionate different exRNA-nanocarriers at high throughput, yield, and purity from various biofluids (plasma, urine, and saliva).

Our platform called the Continuous Isoelectric Fractionation (CIF) device consists of two modules of microfluidic chips: (i) pH gradient chip and (ii) separation chip. The pH gradient chip incorporates a pair of bipolar membranes, which splits water into H_3O^+ and OH^- ions under optimized reverse bias voltage due to Wien effect. The separated H_3O^+ and OH^- ions are separated and stabilized by flow (~12 ml/hour), zero-flux Boltzmann distributions of the inert buffer ions as well as rapid and reversible water dissociation reaction to form linear pH gradient (pH 3 to 11) without using commercial ampholytes. Depending on the application, the desired portion of the effluent from the pH gradient chip is then injected downstream into a separation chip(s) to produce a high-resolution pH gradient where the nanocarrier mixture is injected (~3ml/hour) and separated based on their charge (isoelectric point). To facilitate desired pH transfer from the pH gradient chip to the separation chip, a machine learning procedure is developed. For quantitative assessments, the effluents collected from all the outlets of the chip were qualitatively analyzed by various methods such as ELISA, gel electrophoresis, zeta potential, RT-qPCR, and TEM images.

We optimize the CIF technology by fractionating various combinations of binary mixtures of exRNA nanocarriers (EVs, LLPs, and RNPs) spiked in 1x PBS buffer (yield >80% and purity >90%). Its performance is then evaluated to fractionate RNPs from other nanocarriers in several biofluids, including plasma, urine, and saliva samples. Comprehensive, high-purity (plasma: >93%, urine: >95% and saliva: >97%), high-yield (plasma: >78%, urine: >87% and saliva: >96%), and probe-free isolation of ribonucleoproteins in 0.75 ml samples of various biofluids in 30 minutes is demonstrated, significantly outperforming affinity-based and highly biased gold standards having low yield and day-long protocols. The optimized technique has a resolution of 0.3 ΔpI, sufficient to separate many nanocarriers and even subclasses of nanocarriers. This device represents a significant advancement as it overcomes the various limitations described earlier of commonly utilized traditional isolation technologies including ultracentrifugation, immunocapture, and ultrafiltration techniques.

Collectively, this scalable device enables an exciting approach to early disease detection and may provide an effective technique to validate as well as identify new disease biomarkers in body fluids. Additionally, we envision CIF platform's future application in exploring the heterogeneity of EVs such as fractionating different EVs types with cargoes derived from cancer cells (e.g. GPC-1, Active EGFR, AR-V7). It will require a fine pH gradient, which will likely require multiple separation devices to achieve sufficient pH resolution. We also anticipate that this technology can be used to purify other biological nanoparticles including virus vaccines, exosome drug carriers, amyloid-beta aggregates and peptide assemblies.

251f Effects of Flowrate, External Magnetic Field Strength, and Total Number of Cells Fed on Performance of Catch-and-Release Style Magnetic Separators for Red Blood Cells.

Jacob Strayer, Xian Wu, Hyeon Choe, Jeffrey Chalmers, Jenifer Gomez Pastora

Magnetic cell separations are a simple and effective way of sorting red blood cells based on the cell's oxygenation equilibrium profile. Magnetic techniques can be especially useful in treating sickle cell disease (SCD) because of the uniquely different oxygenation equilibrium caused by the disease. Further understanding of the relationships between variables governing the performance of these separators is needed. The magnetic separation in a catch-and-release style separator is hypothesized to be affected by the balance between magnetic forces holding the cells in the separator to the hydrodynamic force pulling the cells off the separators. In this study, we investigated the effect of flow rate in packed column magnetic separators exposed to different external magnetic field strengths.

251h Process Intensification in Biofuels Manufacturing – Characterization of Liquid-Liquid Extraction Behavior of Biofuel Components

Gnanaselvan Gnanasekaran, Lucas Stolp, Hua-jiang Huang, Shri Ramaswamy

Biofuels including ethanol manufacturing from first- and second- generation feedstocks are an important endeavor in addressing the sustainability goals in the U.S. Despite the advantages of biofuel as a blend with gasoline, there continue to remain opportunities for improving the cost and energy performance, water use and environmental impacts of biofuels manufacturing.

One of the significant contributors to energy consumption during biofuel manufacturing is the process of distillation. Alternative separation technologies including absorption, liquid-liquid extraction, membrane separation and membrane solvent extraction have been explored. Membrane solvent extraction offers significant advantages compared to conventional liquid-liquid extraction including eliminating the formation of emulsions, ability to use solvents that have potential toxicity to microorganisms at higher concentrations, ease of separation by using high boiling solvents. Solvent extraction from fermentation broth or liquid (solvent) - liquid (aqueous broth) extraction of the biofuel components is an integral part of membrane solvent extraction. Understanding the equilibrium extraction characteristics and partition coefficients of biofuel components is an important first step in developing advanced separation technologies such as membrane solvent extraction.

Here we present experimental characterization and analysis of aqueous and organic phases and equilibrium partitioning during liquid-liquid extraction in single and multi-component mixtures and their comparison to literature and theoretical predictions using thermodynamic modeling. The concentration and equilibrium partitioning characteristics will be helpful in the development of advanced separation technologies and improving the cost and energy performance of biofuels manufacturing.

329a Feedstock-Agnostic Reductive Catalytic Fractionation in Alcohol and Alcohol-Water Mixtures

Jun Hee Jang, Ana Colaco Morais, Yuriy Román-Leshkov, Gregg Beckham

Feedstock flexibility is a key challenge for implementation of biomass conversion technologies at industrial scale due to the limited access to a single biomass feedstock with year-round availability at low cost. Reductive catalytic fractionation (RCF) is a promising lignin-first biorefining process, wherein lignin is selectively extracted from native plant cell wall with a polar protic solvent and depolymerized to stable aromatic monomers through reductive catalysis. In reports to date, RCF has been demonstrated to be applicable to a wide range of lignocellulosic feedstocks including hardwoods, softwoods, grasses, and agricultural residues, suggesting the potential for this process technology to be feedstock agnostic. To that end, here we study the impact of feedstock type on RCF performance to provide a self-consistent, comprehensive assessment regarding the feasibility of feedstock-agnostic RCF. Specifically, RCF was performed with poplar (hardwood), switchgrass (herbaceous monocot), corn stover (herbaceous monocot), and pine (softwood) at varying solvent compositions (alcohol and alcohol/water) in both batch and flow-through (FT) reactors.

Batch and FT reactions performed with methanol revealed delignification values and monomer yields greater than 65 wt% and 25 wt%, respectively, for the tested hardwood and grasses despite their inherent differences. RCF of pine exhibited lower lignin extraction efficiency (<40 wt%), but the high content of lignin in pine enabled a similar lignin oil yield on a biomass basis relative to other feedstocks. In the presence of water as a co-solvent, delignification values increased to greater than 78 wt% regardless of the tested feedstock. The comparable delignification values, together with similar lignin oil and carbohydrate yields observed across the tested feedstocks, suggest the potential of RCF with an alcohol/water mixture as a feedstock-agnostic process.

329b Depolymerization of Lignin for the Production of Phenolic Monomers

Suping Zhang

Lignin is an abundant aromatic polymer in the world, and the depolymerization of lignin into phenolic monomers is one of the main methods for its efficient utilization. However, due to its complex structure and low reactivity, achieving directional depolymerization and improving product yield and selectivity is considered to be one of the most important challenges in lignin applications.

In this paper, a metal-supported molybdenum carbide catalyst was developed to depolymerize lignin for the selective production of high value-added phenolic monomers. The characterization results of the catalysts showed that the Mo_2C was successfully synthesized by impregnation coupled with carburizing reaction method. And the introduction of metal could not only promote the formation of β-Mo_2C active catalytic center, but also improved the specific surface area and structure of the catalysts. Under the optimized conditions (260 °C, 4 h, 20% methanol addition), the Ni-Fe supported molybdenum carbide catalyst can depolymerize lignin to obtain 89.56% liquid yield and 35.53% phenolic monomers, and among them the yield of 4-ethylphenol reached 14.77%. The results indicated that the efficient and selective depolymerization of lignin under mild conditions were realized. The depolymerization mechanism was studied by techniques such as NMR and HPLC-MS. The results showed that during the depolymerization of lignin, unstable intermediates were formed first. Afterwards methanol and active hydrogen would react with these intermediates to convert them into liquid products, thereby inhibiting the formation of coke. At the same time, Ni and Fe metals in the catalyst can inhibit the conversion of 4-ethylphenol to phenol by controlling the hydrogenolysis reaction of the lignin side chain, thus achieving the purpose of selectively production of phenolic monomers.

Finally, the depolymerization mechanism of lignin was studied by the density functional calculation (DFT). The results showed that the bimetallic NiFe(111)-Mo_2C(001) catalyst had better adsorption effect for the model compound than Mo_2C(001), and can reduce the bond energy by stretching the bond length of β-O-4, then promote the depolymerization reaction. At the same time, the catalysts had a lower adsorption energy for the partially hydrodeoxygenation products, which was beneficial for their rapid desorption from the catalysts surface, and then removed from the reaction. On the other hand, the catalysts had strong adsorption to the products containing C=O, which can facilitate the formation of target products by hydrodeoxygenation reaction. The study of the β-O-4 cleavage mechanism showed that the reaction energy barrier for H-adduct at the β position first followed by the β-O-4 cleavage was lower than that of the one-step hydrogenolysis of the β-O-4 bond, which was also beneficial for the hydrodeoxygenation of the products. For MME, the total reaction energy barrier of the depolymerization starting with the hydrogenation of C=O was higher than that of starting with the cleavage of β-O-4, indicating that the cleavage of β-O-4 prior to the hydrodeoxygenation of C=O is the most reasonable reaction route, which also verifies the experimental results from MME depolymerization.

329c Revealing Mechanistic Insights into Lignin Decomposition through Pygc-MS Analysis

Heather LeClerc, Jeffrey Page, Zachary Manfredi, Chase McGee, Katelyn Honegger, Andrew R Teixeira, Julia A. Valla, Michael T. Timko

100 million tons of lignin waste is generated from the pulp and paper industries annually.[1] Lignin is the aromatic macromolecule of biomass composing between 15 – 30% of plant matter.[2,3] Due to its complex aromatic structure, lignin has historically been difficult to characterizend degrade without severe reaction conditions. If we are able to apply precision molecular scissors to the macrostructure of lignin, then we can transform lignin from a million ton waste problem to a valuable resource capable of producing synthetic wood materials with enhanced strength and flexibility and aromatic specialty chemicals traditionally derived from fossil fuels.

Lignin serves as an important renewable biomolecule, yet differences in its fractionation method and native structure can result in varying degrees of depolymerization and product yields. This work aims to understand the effect that fractionation can have on decomposition pathways through an analysis of Kraft and CELF lignins. Hydrothermal liquefaction (HTL) is a promising technique to produce a carbon-rich biocrude, however, it is not suitable for understanding the complex degradation pathways of lignin due to long heat-up and cool-down timescales. HTL of lignin has shown increased biocrude yields from hardwood lignin, yet monomer yields remain below 20 mg/g lignin. For this reason, fast pyrolysis coupled with a GC-MS was utilized to understand the effects of wood type and fractionation method of lignin.

In this way, CELF lignin has been shown to result in increased monomer production compared to Kraft-highlighting decreased β-O-4 linkages in CELF lignin, as identified with NMR. By using step-wise pyGC, the temperature-dependent decomposition products were determined from Kraft and CELF fractionated soft and hardwoods. The total liquid carbon yield from CELF softwood lignin increases steadily with increasing pyrolysis temperature, whereas Kraft softwood experiences a maximum liquid carbon yield at 400 °C before a 10% decrease at 500 °C. In addition to pyrolysis carbon yields, the initial lignin as well as the resultant char were both analyzed with ATR FT-IR to determine the changes in bulk functional groups after degradation. This work will reveal compositional-dependent pathways for the thermal degradation of pre-fractionated lignin.

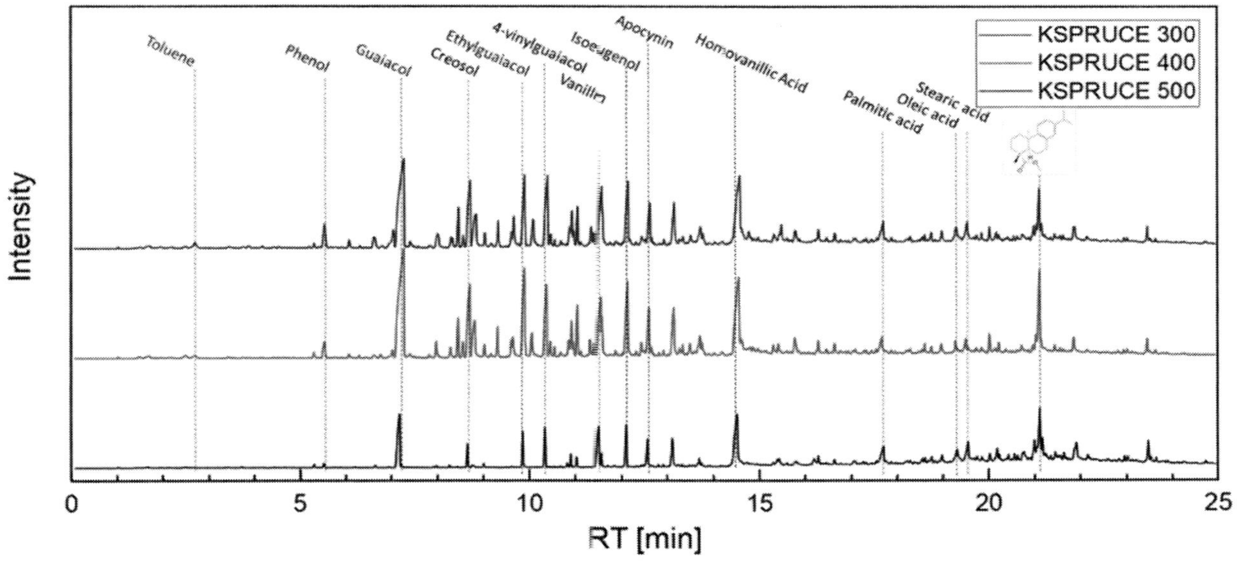

329d Mixed Metal Oxide Catalysts for Microwave-Assisted Kraft Lignin Depolymerization to Produce Phenolic Monomers

Kirtika Kohli, Ravindra Prajapati, Jaemin Kim, Brajendra (BK) Sharma

For the economic viability of biorefineries, the efficient valorization of biomass-derived wastes, lignin, is important. Lignin, which accounts for 15-30 % of the available renewable carbon can be processed to produce value-added phenolic chemicals. In this study, microwave-assisted depolymerization of a Kraft lignin in methanol / formic acid (FA) solvent to produce high-value added phenolic monomers is investigated, in which methanol is a solvent, while FA acts as an acid catalyst and in-situ hydrogen donor. The highest lignin conversion of 78 % was achieved at 140°C and a FA-to-lignin mass ratio of 4 after 20 mins reaction time. Acid-catalyzed cleavage of the linkages in lignin occurs in the presence of FA, which results in the formation of a liquid product. Further, the catalytic depolymerization of Kraft lignin with added methanol / FA solvent system was conducted in the presence of mixed metal oxides to maximize liquid product yield and minimize the solid residue formation. Three different mixed metal oxide catalysts were prepared using hydrotalcite-structured materials i.e., Mg_3AlCO_3, Mg_3AlSO_4, and Mg_3AlCl. Mixed metal oxide-Mg_3AlCl catalyst having a particle size of 50 nm was found to be effective for Kraft lignin depolymerization. The effects of the catalyst composition, catalyst amount, and reaction conditions on the lignin conversion and product yields were explored. The highest lignin conversion of 89 % (with 81% bio-oil yield) was obtained at an optimized reaction temperature of 100°C with a catalyst-to-lignin ratio of 0.25, FA-to-lignin mass ratio of 4, and at a 20 min reaction time. The depolymerized liquid product contains mainly G-type phenolic monomers (~60 %) in the presence of mixed metal oxide-Mg_3AlCl. This work provides an efficient approach for the production of value-added phenolic compounds from lignin by microwave heating in the presence of mixed metal oxides and without external hydrogen.

329e Use of Biorenewable Alcohols in the Production of Lignin for High-Value Applications

Biljana Bujanovic

Lignin has been explored as an alternative to synthetic polymers for various applications ranging from additives for asphalt, thermosets and thermoplastics to hydrogels for drug delivery. Intrinsic lignin properties, such as UV-absorption, radical-quenching, and antimicrobial properties, offer performance benefits to these materials. In addition, low-molecular weight aromatic and phenolic compounds produced from lignin may be used to replace fossil-carbon chemicals. However, inconsistent physicochemical properties and contaminants, such as carbohydrates and inorganics, along with a more condensed nature of technical lignins in comparison to that of lignin in situ are notoriously recognized as obstacles hindering broader use of lignin in high-value applications. This study aims to explore lignins produced after organosolv delignification conducted on both aspen and hydrothermally pretreated aspen using select biorenewable alcohols, specifically: low-boiling point alcohol, ethanol, and high-boiling-point alcohols, ethylene glycol, propylene glycol, 1,4-butanediol, and tetrahydrofurfuryl alcohol. These alcohols are selected based on their solubility parameters placing them relatively close to lignin in the Hansen space. Aspen lignins recovered from organosolv liquors are thoroughly characterized and compared to aspen MWL, and commercially available hardwood lignins, Alcell and kraft. Our goal is to assess the ability of these alcohols, applied individually or in conjunction with hydrothermal pretreatment, to selectively remove lignin from aspenwood chips and, produce lignin of improved properties, including purity and preserved chemical structure.

329g Chemical Modification of Lignin: Increasing Molecular Weight and Polarity Control of Lignin Via Esterification

Oreoluwa Agede, Mark Thies

As a key component of lignocellulosic biomass, lignin is an abundant polymer, notable for its relatively high char yield, low cost, and its functionality making lignin an excellent candidate for downstream applications such as carbon fibers and polymeric blends. However, lignin is still underutilized due to some of its undesirable properties such as poor compatibility with polymers, low reactivity, and wide distribution of molecular weight. To improve lignin's compatibility with commonly available polymers such as polyethylene and polystyrene, there is a need for polarity control. Furthermore, the processing and mechanical properties of lignin-based products such as carbon fibers and composites are dependent on the molecular weight (MW) and glass transition temperature (T_g) of lignin. Recent work in our group has shown that increases in MW and glass transition temperature of lignin are beneficial to the processing and mechanical properties (tensile strength and modulus) of lignin-based carbon fibers.

To improve the performance of lignin, this biopolymer has been subjected to a variety of techniques that involves physical modification such as fractionation, or chemical modification such as functionalization of its hydroxyl groups and depolymerization. Physical modification is commonly used for obtaining molecular weight control, however the desired goal set by the Department of Energy (DOE) for lignin-based carbon fibers (tensile strength and modulus of 1.7 and 172 GPa respectively) has yet to be obtained. There has been little to no work utilizing chemical modification to increase the molecular weight of lignin, but chemical modification has been shown to yield significant improvement in flowability with polymers, and to improve the durability and uniformity of the final products. Therefore, to be utilized for carbon fibers and polymeric blends, it would be of great advantage to chemically modify lignin to target increasing molecular weight, improving flowability with other polymers, and obtaining better polarity control.

This work will focus on interlinking lignin molecules by functionalization of the hydroxyl group present in lignin with di-functional or multi-functional molecules while simultaneously interlink lignin molecules. By modifying and reducing the hydroxyl group, and thus decreasing the polarity of lignin, its compatibility with non-polar polymers will also increase making it a better alternative for polymeric blends and composites. Using this technique, we will be able to improve the properties of lignin by increasing molecular weight and glass transition temperature while also reducing polarity.

This work is investigating the effects of interlinking, reaction conditions, structure of the interlinking agent, and the extent of interlinking on the performance of lignin and the subsequent products.

329h A New Process for Recovering High Quality Lignins from Agricultural Black Liquors

Bronson Lynn, Zachary Pamukcoglu, Mark C. Thies

Utilization of all the components of lignocellulosic biomass is essential for biorefineries to compete economically with their petroleum counterparts and establish themselves as a cornerstone of the green economy. Agricultural residues left over from annually harvested food crops, along with fast-growing energy crops, represent a vast resource that can be tapped for cellulosic bioethanol or bioproducts, but progress has been limited by the profitability of these biorefineries. The lignin byproduct from these biorefineries is key to resolving this issue, as it is the second most abundant component in the biomass and offers aromatic character that the carbohydrates lack.

The separation and subsequent extraction of the carbohydrates and lignin in the biomass is usually achieved through either dilute acid pretreatment or alkaline pretreatment, with the latter being favored here for its tendency to leave the cellulose and the lignin less degraded. However, little work has been done to develop economically and environmentally friendly techniques to recover lignin from the lignin-containing byproduct stream (known as black liquor). Instead, most techniques today for recovering lignin from agricultural black liquors involve energy-intensive centrifugation steps and/or extensive water washing. Considering lignin is intended to be part of the solution to the sustainability problem, using expensive or polluting techniques on what is supposed to be a low-value green material defeats the purpose lignin was originally intended to serve.

In this work, we have developed a process to recover lignins directly from the alkaline black liquors of multiple agricultural feedstocks (e.g., corn stover and wheat straw). The process requires no washing or centrifugation and only one filtration step. This new recovery process, when used in conjunction with the Aqueous Lignin Purification using Hot Agents (or ALPHA) process, provides the missing link to directly convert agricultural black liquors into high quality lignins of <0.25% sugar and <0.5% ash with molecular weights differing by an order of magnitude. The ALPHA process uses only simple green solvent systems like ethanol/water and acetone/water at low solvent-to-feed ratios (<6:1 S:F) to accomplish simultaneous fractionation and purification. In summary, this new recovery technique offers an economical and sustainable alternative to traditional recovery and purification/fractionation techniques currently being applied to the alkaline liquors of agricultural residue streams.

329i Polymer Products from Lignin through De-Aromatization and COOH Functionalization

Michael Kent, Daniella Martinez, Qi Wang, Mitra Ganewatta, JiHyuan Hwang, Chuanbing Tang

Background and motivation. Valorizing lignin has the potential to significantly improve the economics of biomass conversion technologies. Lignin comprises 15-30% of lignocellulosic biomass and is currently burned for its heating value within a biorefinery. Generating greater value out of this carbon is critical for the economic viability of producing fuels and chemicals from lignocellulosic biorefineries.

We report a method to convert lignin into valuable polymer products. Compared to depolymerizing lignin, functionalizing lignin has the advantage of yielding a product directly without requiring further upgrading. We use an inexpensive aqueous oxidative process near room temperature to open the aromatic rings within the lignin structure and generate carboxylic acid groups. This polymeric polyacid material functions as a commercial agricultural dispersant, micronutrient complexation agent, or water-absorbent material. No separation processes are required and all of the lignin is converted to product.

Methods. Chemical modification and functionalization of lignin was achieved using a chelator-mediated Fenton (CMF) reaction. A range of samples were prepared from kraft lignin and also from a lightly-processed near-native lignin from a lignocellulosic biorefinery (NREL's DMR-EH lignin). The H_2O_2 requirement for these samples, expressed as mass of H_2O_2 (pure basis) per mass of lignin, ranged from 1 to 0.25 depending upon the extent of functionalization. Upon optimizing reactant concentrations and pH, the COOH content ranged up to ~2.5 mmol/g. Further, under certain conditions the molecular weight was increased by a factor of 10. Increasing the molecular weight is important for certain applications. The samples were tested for performance as a dispersant, for micronutrient complexation, and for water purification. Further, crosslinked hydrogels were generated from the functionalized lignin using acrylamide and N, N'-methylenebisacrylamide after imidization of the acid groups on the lignin, as reported in Giant 2022, 10, 100106.

Results. A set of samples were prepared from each lignin source that differ in COOH content and molecular weight. These samples were evaluated for i) their performance in formulations to disperse the herbicide mesotrione, insoluble metal oxides, and soluble metal salts, ii) their ability to bind divalent heavy metal ions (Pb(II), Co(II), Ni(II) Cu(II), and Zn(II)), and iii) for formation of hydrogels and water-absorption performance. The results are summarized in Figure 1 and will be discussed in more detail in the presentation. This chart shows that many of the functionalized lignin samples derived from kraft lignin meet performance specifications for all applications except dispersing soluble metal salts, whereas the samples derived from biorefinery lignin only met performance specifications for hydrogels. The poor performance of the samples derived from biorefinery lignin in the other applications was due to the high content of polysaccharides.

Implications. This work demonstrates that a simple low-cost aqueous oxidation treatment can be used to tailor both chemical functionality and molecular weight of lignin, and thereby generates useful polymer products from lignins.

Samples derived from raw kraft lignin

sample	% aromat. IR	COOH (mmol/g) titration	aromat. OH (mmol/g) titration	Mw* SEC
kraft lignin (untreated)	100	1.03	2.47	1,960
R1 (different reaction conditions)	68	1.16	1.55	1,700
R2 (different reaction conditions)	25	1.50	1.29	2,560
R3 (different reaction conditions)	34	1.35	1.08	16,000
R3B (different reaction conditions)	18	2.66	1.05	4,800
R4A (different reaction conditions)	24	1.82	1.14	8100

*measured by SEC using UV 210 nm and 1 mM PO4 pH 9

Samples derived from raw biorefinery lignin

sample	% aromatic IR	COOH (mmol/g) titration	aromatic OH (mmol/g) titration	Mw* SEC
BL** (untreated)	100	0.36	0.61	65,000
BL-BCD (different reaction conditions)	------	1.68	0.83	9,700
BL-CMF1 (different reaction conditions)	0.57	2.30	0.76	60,000
BL-CMF2** (different reaction conditions)	0.04	1.70	0.72	56,000
BL-CMF3A** (different reaction conditions)	0	1.23	0.70	104,000
BL-CMF3B (different reaction conditions)	0	2.55	0.88	20,000
BL-BCD-CMF1 (different reaction conditions)	------	1.47	0.60	21,000
BL-BCD-CMF2 (different reaction conditions)	------	1.92	0.68	16,000
BL-BCD-CMF3 (different reaction conditions)	------	1.83	0.54	34,000

*measured by SEC using UV 210 nm and 1 mM PO4 pH 9
**highly insoluble at pH 12 (analyzed soluble portion)

performance results

	kraft lignin	R1	R2	R3	R3B	R4A
mesotrione dispersion		✓	✓	✓		
insoluble metal oxide dispersion		✓	✓	✓	✓	✓
soluble metal salt dispersion						
water purification Pb(II) Co(II) Ni(II) Cu(II) Zn(II)			✓		✓	
water absorption		✓	✓	✓	✓	

performance results

	BL	BL-BCD	BL-CMF1	BL-CMF2	BL-CMF3A	BL-CMF3B	BL-BCD-CMF1	BL-BCD-CMF2	BL-BCD-CMF3
mesotrione dispersion									
insoluble metal oxide dispersion									
soluble metal salt dispersion									
water purification Pb(II) Co(II) Ni(II) Cu(II) Zn(II)									
water absorption		✓	✓						✓

✓ = meets minimum performance metrics

157a "Black a New Green" – Innovation in Sustainable Manufacturing

Amar K. Mohanty

Waste biomass on pyrolysis creates pyrolyzed biocarbon ("Black a new green") that finds exciting opportunities recently in sustainable manufacturing from high barrier packaging to light-weight automotive parts. The biocarbon (or biochar) is the solid carbonaceous material made through thermochemical conversion. This biocarbon material is mainly responsible for the characteristic dark colour and high carbon content of terra preta ('black soil', a.k.a. 'Amazonian dark earth') found in the Amazon basin. This technique of using charred organic matter to improve the soil fertility originated at least 7000 years ago and has been hailed as a model for sustainable agriculture. Carbon allotropes have been recognized as the most versatile material in the modern science and technology (materials, energy and environmental remediation applications) due to their structural and morphological diversity influenced by the ability of carbon to form different hybridizations. Thus the discovery of biocarbon in materials uses is quite exciting and challenging. We have developed light-weight biocomposites from sustainable biocarbon filler and showed how these fillers can replace talc and short glass fibre in emerging light-weight injection moulded automotive parts. Advance biocarbon on reinforcement with recycled engineering plastic helps in reducing the flammability of the composites significantly that can be used in demanding electric vehicle (EV) parts. In another ground breaking discovery we showed how biocabon materials can act as oxygen scavenger in bioplastic in the design and engineering of high barrier packaging material for food packaging uses. This presentation will highlight innovation and commercialization of versatile biocarbon in materials applications from eco-friendly and light-weight parts including EV parts to high barrier plastic for sustainable packaging.

157g Catalytic Upgrading of Carbohydrates in Paper Sludge to Sustainable Aviation Fuels

Sunkyu Park

The primary objective of this project is to develop a technology for converting the carbohydrates in paper sludge, a wet organic industrial-waste stream, into a sustainable aviation fuel (SAF), both economically and sustainably. US-DOE has identified wet organic waste streams as valuable potential feedstocks for the bioeconomy. This project will develop an integrated process using wet paper sludge where no drying of the feedstock is needed. The process includes (a) ash removal from paper sludge, (b) enzymatic hydrolysis of carbohydrates to monosaccharides, (c) dehydration of pentoses and hexoses to furfural and HMF, (d) aldol condensation of furans with ketones to intermediates having molecules with 10-14 carbons, (e) hydrodeoxygenation of the intermediates to paraffins with excellent properties for blending in jet, and (f) robust techno-economic analysis (TEA) and life cycle assessment (LCA) to focus research on developing cost-effective routes to address key cost barriers, and ensure the sustainability of the process. The developed process will be validated at a relevant scale to produce sufficient HC biofuel for fuel property testing.

The team produced a hydrocarbon fuel from paper sludge for aviation fuel testing and its performance was comparable to a commercial jet fuel. During the presentation, the process optimization and trade-off analysis between TEA and LCA will be explained in details.

This project has been supported by US Department of Energy (DOE) and plans to be completed in 2023.

157c Sustainable Bio-Based PU Materials from Lignin and Other Biomolecules

Ning Yan

Polyurethanes (PU) are versatile polymeric materials available with a wide range of properties and forms. However, conventional PU materials are petrochemical products derived from fossil fuel resources. there is a growing interest in exploring renewable feedstock as building blocks to increase sustainability. In this presentation, several novel green chemical pathways for making bio-based PU systems containing different types of biomolecule precursors are highlighted. Lignin was explored as bio-based polyols to make lignin-based PU foams capable of rapid clean-up of crude oil spills with outstanding adsorption capacities, while these bio-based PU foams could also be degraded in alkaline environments after usage to reduce secondary pollution concerns. A new strategy of catalyst-free direct synthesis of covalent adaptable network polyurethanes from lignin with editable shape memory effect was also successfully developed. Moreover, a new family of fully bio-based shape memory PU materials was obtained by synthesizing a bio-based diisocyanate from abietic acid and a novel bio-based chain extender derived from vanillin and alaninol amino acid. These bio-based PUs exhibited excellent thermal-induced functional shape memory responses combined with a good degradability. The research results included here showcased excellent promises of using lignin and other biomolecules to produce bio-based PU materials as more sustainable replacements of fossil fuel derived products.

157d Green-solvent Assisted Processing (GSAP) of Polylactic Acid Nanocomposites with Hybrid Cellulose Fibers

Lu Wang, Griffin Miller

Hybrid fiber-reinforced polymer composites (HFRPCs) are materials in which a particular polymer matrix is reinforced with more than one reinforcing materials. HFRPCs are multifunctional materials that bear more than one characteristic benefit. In this study, hybrid cellulose fibers, e.g., pulp fiber (PF) and cellulose nanofibril (CNF), were used for reinforcing polylactic acid (PLA), a popular material for extrusion-based additive manufacturing (EAM), because of their relatively low-cost, abundance, and renewability. The HFRPCs were produced with varying ratios between PF and CNF at a fixed total fiber loading of 20 wt.% to investigate the effect of the fiber composition on the composite tensile and physical performances. HFRPCs displayed improved CNF dispersion, increased tensile performances, and heat deflection temperature, compared to single-reinforcement FRPCs. In addition, a demonstrative print of EAM of a recycle bin using one of the developed formulations was implemented.

157e Optimization of Biobased and PCR Compounds Utilizing AI/ML Technology

Arash Kiani

Biobased polymers and PCR compounds are hot topics these days for many industries such as Automotive, Appliance and Industrial. Many global companies are heavily invested and engaged in developing products based on bio polymer and/or PCR compounds. Difficulties in processing biopolymers and sorting and cleaning PCR compounds adds new challenges and most often takes large capital investment and spending extra manpower to overcome the problems. This takes a tremendous toll on the financial health of a company. AI and ML technology is the panacea. This presentation discusses the advancements Alterra has made in the last 5 years in AI and ML to optimize compounding process for the biobased and the PCR compounds.

157f Achieving Sustainability: A Career in Automotive

Deborah Mielewski

The world is rapidly changing and the evidence of global degradation by humans is hard to deny. The topics of sustainability and circularity are at the forefront of discussions in academics and the media, and almost every company has made commitments to zero carbon, zero waste, using recycled materials, and increased sustainable materials content. But are these efforts real and are they audacious enough to help?

I spent the entirety of my career working for one company and spent the past 25+ years pushing to implement sustainable initiatives from within. My learnings from the laboratory drove me to a sobering conclusion--that many lower-impact materials are in fact superior to those industry has used for the past 75 years. They could also be produced at competitive or reduced cost, and many were significantly lighter in weight. So why is it so difficult for meaningful change to happen?

Reviewing our teams' previous successes in implementing over 15 industry-first, sustainable materials and the roadblocks and barriers that were overcome to get there can help us understand how we might accelerate progress toward greener initiatives. The role of industrial partnerships and the environmental benefits that can be achieved through unlikely collaborations will also be discussed.

Last, I will present what I envision to be some of the next big environmental challenges that have yet to be adequately addressed, and how re-thinking the way we've always done things can bring huge benefits for the future of our planet.

562b Computational Study on the Catalytic Reductive Depolymerization Mechanism of a *B*-O-4 Lignin Dimer Model Compound in Subcritical Methanol

Panuwat Watthaisong, Subrata Kumar Kundu, Andreas Heyden

Lignin is one of the three primary components of lignocellulosic biomass, which is generally composed of three phenylpropane units, i.e., p-hydroxyphenyl (H-type), guaiacyl (G-type), and syringyl (S-type) units which are interconnected by several ether and C–C linkages. Traditionally, lignin has been treated as a waste product in the pulp and paper industry, where it is burned to supply energy and recover pulping chemicals in the operation of paper mills. To utilize lignin more efficiently, reductive catalytic fractionation (RCF) has emerged as a promising way to achieve the conversion of lignin to value-added chemicals. This approach combines solvolytic extraction of lignin from holocellulose (cellulose and hemicellulose) and reductive stabilization of reactive intermediates, leading directly to a depolymerized lignin oil. Conventional RCF was carried out in batch reactors where biomass, a reducing catalyst, and a hydrogen donor are combined in polar protic solvents. However, it has been reported that the physical mixing of the biomass and catalyst complicates kinetics studies. Also, lignin extraction is solely dependent on the solvolytic conditions and does not require a redox-active catalyst. On the other hand, reductive stabilization of the unsaturated fragments is exclusively governed by the redox-active catalyst. Therefore, several reactor configurations have been applied to physically separate the solvolytic lignin extraction-depolymerization from the catalytic hydrogenolysis-stabilization step for RCF processes, including conducting solvolysis and hydrogenolysis as separate batch reactions or in tandem flow-through reactors. Polar-protic solvents such as methanol are usually used at high temperatures (up to 250 °C) in combination with a heterogeneous catalyst (e.g., Pt/C, Ru/C, or Pd/C), to obtain high-yield lignin oil that is rich in aromatic monomers with saturated side chains.

Computational studies can provide a mechanistic understanding of the reaction pathways and the effect of solvation in RCF. In our work, density functional theory (DFT) calculations were used to develop a mechanistic understanding of the underlying chemistry of two major processes in the RCF approach, solvolysis and stabilization. To better understand solvolysis, guaiacylglycerol-beta-guaiacyl ether (GGE) was employed as the model compound of lignin to understand the relative reactivity and chemical pathways that a β-O-4 bond will undergo in subcritical methanol, producing coniferyl alcohol and guaiacol. A key objective is to investigate the role of subcritical methanol in this reaction, which acts as a hydrogen donor, solvent, and reactant simultaneously. Both microsolvation and implicit solvation approaches have been used in this case to explore the effects of subcritical methanol. Concerted retro–ene and Maccoll elimination reactions (which are suggested to be the dominant pathway for typical pyrolysis temperatures between 500 and 600 °C) have been studied to evaluate the possibility of the solvent promoting these reactions in subcritical conditions (~230 °C). Our calculations show that the depolymerization of the GGE molecule through a retro-ene fragmentation reaction is kinetically and thermodynamically more favorable compared to Maccoll elimination reactions. The reactions start with the intramolecular hydrogen transfer, which is also the rate determining step. The free energy barriers of this process are 245 kJ/mol in the gas phase and 212 kJ/mol with the inclusion of implicit solvent. Even though the implicit solvent approach reduces the reaction barrier by 33 kJ/mol, the barrier remains high for a reaction to occur at a temperature of 500 K. However, a potential depolymerization mechanism is found using the microsolvation approach. We found that the reaction proceeds through a quinone intermediate, which is assisted by intermolecular hydrogen transfer from phenolic group of GGE molecule and two explicit methanol molecules. Then, one of these methanol molecules donates two hydrogens to activate the β-O-4 bond cleavage of the GGE molecule. This reaction pathway agrees well with a number of experimental observations including the retention of α and β hydrogen, the formation of formaldehyde in the reaction mixtures, and the significance of the phenolic group in the lignin model compound. It can be concluded from this result that the methanol molecule plays a vital role as hydrogen donor and acceptor that facilitate the depolymerization of lignin. Moreover, the calculations suggest that, in some cases, the implicit solvent might not be sufficient for modeling the reaction in the liquid phase and one or more explicit solvent molecules might have to be included.

In addition to the solvolysis reaction, we have also studied catalytic hydrodeoxygenation/hydrogenation reactions of coniferyl alcohol over the Pd(111), Pt(111), and Ru(0001) catalyst surfaces, a process called stabilization. Coniferyl alcohol represents the reaction product from solvolysis. In this case, the solvent effect on the rate-determining step has been taken into account using our hybrid QM/MM method, called Explicit Solvation for Metal surfaces (eSMS). Since it captures the true molecular nature of the solvent molecules, eSMS can reliably predict solvation effects on metal surfaces. To conclude, by studying solvolysis and stabilization of a model reactant such as GGE, we hope to be able to provide a comprehensive picture of the lignin conversion process during RCF with the goal of optimization both the reaction conditions and the catalyst.

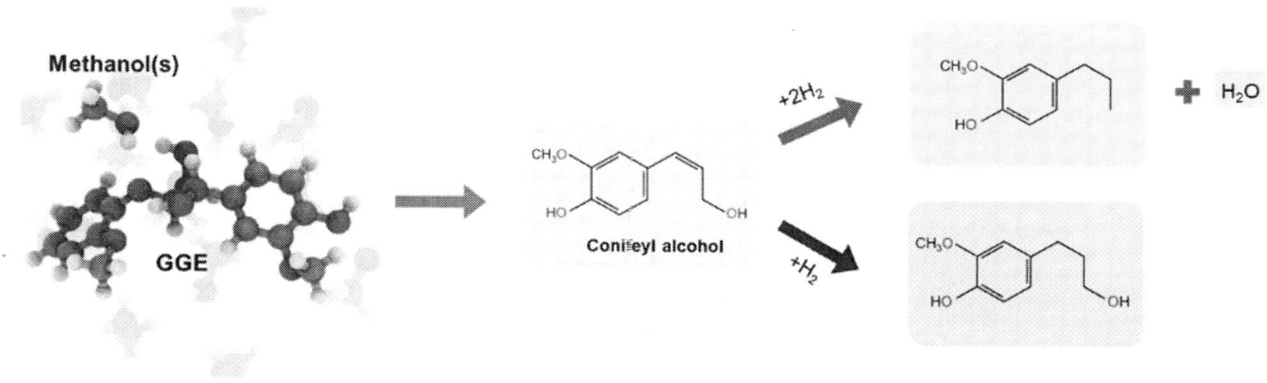

562c Reinforcement Learning Based Control for Biorefining Process Under Uncertainty

Abigael Wahlen, Ji Gao, Caleb Ju, Yongsheng Chen, Guanghui Lan, Zhaohui Tong

Excess waste is a major issue affecting global sustainable development and has led to an influx of pollution and greenhouse gas emissions. Biorefineries are an alternative pathway for waste that are able to convert low-cost biomass into value-adding biproducts. Here in, we present a reinforcement learning based framework to control the uncertainties associated with a centralized biorefinery plant. Anaerobic digestion, a well-known and utilized biorefinery process, is used to show the performance and robustness in production and uncertainty control. Our framework outperforms traditional control strategies in short-term scenarios, achieving fast target tracking, increased precision, and accuracy. When considering long term plant control, with inventory limitations, our framework shows adaptive and robust behavior satisfying downstream demands. We have shown that our reinforcement learning framework is a promising and scalable solution for uncertainty issues in real-world biorefining processes, contributing to the advancement of sustainability and waste management.

562d Application of Discrete-Time Infinite-Dimensional Model for Optimal Control and Estimation of a Continuous Pulp Digester

Lu Zhang, Haneen Arauydah, Stevan Dubljevic

In recent years, the interests of the pulp industry and academia have focused on simultaneously producing paper with superior grades and value-added chemicals, owing to the rise of e-commerce and emerging markets worldwide. This growth has led to a need for improved pulping efficiency to mitigate the negative environmental impact of the paper-making industry. Wood chips are important raw materials in this process, and also valuable resources for lignocellulosic biomass, which can be used as alternative fuels and green materials due to its renewable and sustainable nature [1] [2]. Hence, the development of an accurate dynamic model and reliable controllers for the digester system could have a significant impact on the quality and efficiency of the pulping process.

The digester system continuously cooks and washes wood chips from the chip storage before further refining them in the mill and paper-making operations. However, most of the delignification reaction occurs only after the wood chips and white liquor flow down into the subsequent cooking zone, where the mixture is heated to reaction temperatures achieved by liquor circulation through external heaters [3]. Hence, the temperature in the cooking zone is particularly crucial as it greatly affects the quality of the pulp, including the cooking degree and pulp viscosity. In general, the delignification process can be described by a system of nonlinear partial differential equations (PDEs) [4] [5]. Most of the existing contributions design the controller based on the approximated models which are obtained using model reduction or other approximation methods. Motivated by these considerations, this work aims at developing a reliable controller to optimize the temperature of the delignification reaction for cooking wood chips based on the infinite-dimensional properties of PDEs.

As usual in large chemical processes, measurements of process variables in pulping plants are limited and often noisy. Therefore, in this work, a model predictive controller consisting of a state estimator is designed for continuous pulp digester using infinite-dimensional control theories. The underlying model of interest in a pulp digester is modeled by a system of coupled hyperbolic PDEs and ordinary differential equations [6]. Output measurements are considered with delay due to the possible low sampling rate. The Cayley-Tustin transform is utilized to convert the continuous infinite-dimensional system into a discrete one without spatial discretization or model order reduction [7]. A numerical example is provided to demonstrate the feasibility and applicability of the proposed controller designs. The proposed design can be readily extended to the cases of accounting for multiple objective functions to achieve a good balance between biomass recovery and paper production grade.

References:

[1] D. Gavrilescu. Energy from biomass in pulp and paper mills. *Environmental Engineering & Management Journal (EEMJ)*. 2008 Sep 1;7(5).

[2] J. Kim, S. Pahari, J. Sang-Il Kwon. Modeling Biomass Degradation with Multiscale kMC Simulations. In Energy Systems and Processes: Recent Advances in Design and Control 2023 Mar 17 (pp. 11-1). Melville, New York: AIP Publishing LLC.

[3] S. Bhartiya, P. Dufour, FJ. Doyle III. Fundamental thermal-hydraulic pulp digester model with grade transition. *AIChE journal*. 2003 Feb;49(2):411-25.

[4] HK. Choi, SH. Son, J. Sang-Il Kwon. Inferential model predictive control of continuous pulping under grade transition. *Industrial & Engineering Chemistry Research*. 2021 Feb 24;60(9):3699-710.

[5] L. Ding, T. Gustafsson, A. Johansson. Model parameter estimation of simplified linear models for a continuous paper pulp digester. *Journal of Process Control*. 2007;17(2):115-127.

[6] L. Zhang, J. Xie, S. Dubljevic. Dynamic modeling and model predictive control of a continuous pulp digester. *AIChE Journal*. 2022 Mar;68(3):e17534.

[7] S. Dubljevic, and J.P. Humaloja. Model predictive control for regular linear systems. *Automatica*. 2020;119, p.109066.

562e A Multi-Scale Kinetic Modeling and Optimal Control Strategy for an Effective Lignin Fractionation Process

Juhyeon Kim, Jiae Ryu, Mairui Zhang, Chang Geun Yoo, Joseph Kwon

The utilization of biomass, rather than petroleum-based resources, which are associated with environmental challenges and future uncertainties, has become a pressing concern in recent times [1]. Due to its abundance, biomass has garnered substantial research interests, as the scientific community aims to maximize the profitability of biorefineries. In this regard, the successful utilization of biomass components is essential for the commercialization of renewable and sustainable resources. Lignin, a primary constituent of woody biomasses, is currently underutilized despite its potential as a significant source of natural aromatic chemicals [2]. The complex structure of biomass hinders the efficient utilization of lignin, which is often removed as it restricts cellulose accessibility to fractionation reagents.

The utilization of lignin has been emphasized to improve the overall profitability of the biorefinery industry [3, 4]. Although numerous experimental developments have been made [5], a knowledge gap remains in the comprehensive understanding of lignin dynamics in reaction systems. Existing mathematical models [6], which focus solely on de/repolymerization processes with dissolved lignin chains, cannot fully explain the overall biorefinery dynamics. Moreover, real-time measurements of key parameters such as lignin content and molecular weight distribution are still in their infancy [7] and cannot be performed in full-scale pulp digesters. Therefore, a high-fidelity mathematical model and process control strategy are necessary for feasible biomass processing.

Motivated by this need, we introduce a multiscale model that elaborates on both macroscopic (delignification) and microscopic (de/repolymerization) reactions, based on fractionation experiment using a novel solvent, phenol-4-sulfonic acid (PSA) [8]. We employed a bilayer framework, consisting of ordinary differential equations for delignification and a kinetic Monte Carlo (kMC) algorithm for de/repolymerization. Following the successful development of a data-driven kinetic model, we applied a model predictive control to optimize the reaction, using a soft sensor (*i.e.*, Kalman filter) to estimate the primary variables (lignin content and average molecular weight) from the available data (solution temperature).

We verified our model against experimental data and validated our control framework using a real pulp digester, showcasing the seamless integration of theoretical studies with experiments in our work. Our process control demonstrated low offsets from set points for key variables, including lignin content and the average molecular weight of lignin chains. This study highlights the successful combination of experimental work and simulation, leading to a deeper understanding of lignin dynamics in reaction systems and ultimately enabling a lignin fractionation approach in biorefineries.

Literature cited:

[1] Dodds D.R., & Gross R.A. (2007). Chemicals from Biomass. *Science*, 318, 1250-1251.

[2] Gao C., Li M., Zhu C., Hu Y., Shen T., Li M., Ji X., Lyu G., & Zhuang W. (2021). One-pot depolymerization, demethylation and phenolation of lignin catalyzed by HBr under microwave irradiation for phenolic foam preparation. *Compos. B. Eng.*, 205, 108530.

[3] Shuai L., Amiri M.T., Questell-Santiago Y.M., Heroguel F., Li Y., Kim H., Meilan R., Chapple

C., Ralph J., & Luterbacher J.S. (2016). Formaldehyde stabilization facilitates lignin monomer production during biomass depolymerization. *Science*, 2016, 354, 329-333.

[4] Luo Y., O'Dea R.M., Gupta Y., Chang J., Sadula S., Soh L.P., Robbins A.M., Levia D.F., Vlachos D.G., Epps III T.H., & Ierapetritou M. (2022). A life cycle greenhouse gas model of a yellow poplar forest residue reductive catalytic fractionation biorefinery. *Environ. Eng. Sci.*, 39, 821-833.

[5] Abu-Omar, M.M., Barta, K., Beckham G.H., Luterbacher J.S., Ralph J., Rinaldi R., Roman-Leshkov Y., Samec J.S.M., Sels B.F., & Wang F. (2021). Guidelines for performing lignin-first biorefinering. *Energy Environ. Sci.*, 14, 262-292.

[6] Bawareth B, Marino D.D., Nijhuis T.A., & Wessling M. (2018). Unravelling electrochemical lignin depolymerization. *ACS Sustainable Chem. Eng.*, 6, 7565-7573.

[7] Khalili K.N., de Peinder P., Donkers J., Gosselink R.J., Bruijininex P.C., & Weckhuysen B.M. (2021). Monitoring molecular weight changes during technical lignin depolymerization by operando attenuated total reflectance infrared spectroscopy and chemometries. *ChemSusChem*, 14, 5517-5524.

[8] He D., Wang Y., Yoo C.G., Chen Q.-J., & Yang Q. (2020). The fractionation of woody biomass under mild conditions using bifunctional phenol-4-sulfonic acid as a catalyst and lignin solvent. *Green Chem., 22,* 5414-5422.

562f COSMO Simulations and Experimental Evaluation for the Solvation of Lignin in Deep Eutectic Solvents

Thomas Quaid, Laura Fronchetti Guidugli, Toufiq Reza

Lignin and cellulose are the two most abundant natural polymers with wood being the main global source. Cellulose is a sugar polymer which can be converted into biofuel. Lignin is a complex organic structure that encapsulates the cellulose. Due to the vast abundance of these materials and their propensity for being discarded as waste by industry, there exists a strong argument for using it as a cheap and sustainable source of biofuel. The process in which cellulose is converted to biofuel consists of two main stages, hydrolysis, and fermentation. The combined structure of lignin and cellulose undergo hydrolysis to break the sugar polymers of cellulose into glucose monomers. The glucose is then fermented into an ethanol and water mixture which can be separated to obtain the value product of ethanol. During these two processes the lignin disrupts the reaction times and, in some cases, poisons the process. This causes a lower process efficiency and drives up costs making the marketability of the product less feasible.

In this study, deep eutectic solvents (DES) are evaluated to extract lignin. The benefit of DES is that they are made from the physical mixing of a hydrogen bond donor (HBD) and a hydrogen bond acceptor (HBA), generally made from a quaternary salt and acid/alcohols. In this study, a conductor like screening model (COSMO) is adopted to screen suitable HBA's and HBD's for Lignin extraction based on hydrogen bond properties using a triple-zeta valence polarization (TZVP) basis set. This model is evaluated compared to literature experimental studies to verify integrity of the results. Over 1000 combinations of HBA and HBD spanning type three and 5 DES were evaluated for this task. Novel DES combinations involving tetrabutyl ammonium chloride were predicted to offer significant improvements in lignin solubilization over the commonly studied choline chloride-based DES. Experimental procedures were undertaken to determine the validity of the predictions by COSMO. The experiments supported the findings.

562g Machine Learning-Aided Ionic Liquid Design for PET Degradation

Ji Gao, Wenbo Peng, Jose Perez Martinez, Ethan Slaton, Zhaohui Tong

Every year, more than 350 million tons of plastic waste are generated globally, but only a small percentage (9%) is recycled. The rest of this non-degradable waste ends up in landfills and causes water and land pollution. To address this pressing issue, there is an urgent need to upcycle plastic waste into value-added bioproducts. Polyethylene terephthalate (PET) is one of the largest sources of plastic waste, accounting for more than 20% of total plastic waste production. The conversion of PET to useful bioproducts has been intensively studied, and one promising solution is to use ionic liquids as catalysts for converting PET into Bis(2-Hydroxyethyl) terephthalate (BHET) through glycolysis. However, finding a cost-effective and efficient ionic liquid for PET valorization is still challenging. The vast array of available ionic liquids can make experimental screening and parameter optimization a daunting task toward multiple objectives in performance, cost, and environmental impact. To tackle this issue, we present a data-driven approach using machine learning techniques to efficiently design and screen ionic liquids for PET glycolysis. We have compiled a database of approximately 600 ionic liquids and deep eutectic solvents for PET glycolysis based on literature data. By analyzing a range of reaction information such as catalyst, solvent, temperature, time, and size, our machine learning model predicts the performance of different ionic liquids on PET glycolysis. Through this approach, we identified the best performing and cost-effective ionic liquid, a zinc chloride (ZnCl2)-choline chloride (ChCl) based deep eutectic solvent, which demonstrated promising performance and low cost compared to previous studies, and this finding was validated through the experiments. Overall, this method provides an efficient and systematic way to identify ionic liquids for PET glycolysis and can contribute to the development of sustainable plastic waste management strategies.

562h Kinetic Modeling and Techno-Economic Analysis of a Sugar Production Process Via Hydrolysis of Lignocellulosic Biomass

Poulomi Das, Debangsu Bhattacharyya

There is increasingly more emphasis on the development of sustainable technologies for production of energy, fuel, and chemicals. Among many options of sustainable energy sources, lignocellulosic biomass is the only large-scale sustainable carbon source available[1]. Generation of value-added products requires degradation of three major polymeric components[1] of lignocellulosic biomass, namely, cellulose, hemicellulose, and lignin to relatively smaller oligomeric or monomeric units.

C5 (e.g., xylose) and C6 (e.g., glucose) sugars are two of the most important platform intermedia for lignocellulose conversion., production of which can be viable through the hydrolysis process where the glycosidic linkages between the monosaccharides[2] are broken in presence of a catalyst. To recover C5 and C6 sugars from biomass, one of the main technologies is acid-catalyzed hydrolysis that can be performed in presence of either concentrated acid or dilute acid as the catalyst. Concentrated acid hydrolysis (CAH) method has various advantages over dilute acid hydrolysis (DAH) like lower operating temperature, high sugar yield, and low degradation of sugars during hydrolysis. Although there are several limitations of the CAH such as the costs of large amount of concentrated acid(s), corrosion-resistant reactors, and acid recovery process, the interest in CAH has been renewed recently due to the development of acid recovery technologies, and high flexibility of CAH for efficiently handling various types of biomass feedstocks[3]. Existing kinetic model for CAH have been developed for only synthetic components[4,5] and not for the true biomass. In this work, a kinetic model of the CAH is developed, and kinetic parameters are optimally estimated by using the experimental data. A kinetic model of the DAH is also developed. First-principles dynamic models of the batch reactors for CAH and DAH are developed. These reactors are scaled up to the commercial scale. Models of the separation section are also developed for separation of products from the unconverted reactants. In addition, a model is developed for recovering the acid from hydrolysate and recycling it. An economic model of the plant is developed. Techno-economic analysis and optimization are undertaken for maximizing the net present value that takes into account both capital and operating costs and time value of money. Design variables such as the reactor dimensions are optimized along with optimizing the key operating variables such as residence time, catalyst concentration and reaction temperature. Impact of the biomass composition on optimal design and operating variables is evaluated.

References:

1. Zhou, Z., Liu, D. & Zhao, X. Conversion of lignocellulose to biofuels and chemicals via sugar platform: An updated review on chemistry and mechanisms of acid hydrolysis of lignocellulose. *Renewable and Sustainable Energy Reviews* vol. 146 at https://doi.org/10.1016/j.rser.2021.111169 (2021).
2. Shen, X. J., Huang, P. L., Wen, J. L. & Sun, R. C. A facile method for char elimination during base-catalyzed depolymerization and hydrogenolysis of lignin. *Fuel Process. Technol.* **167**, 491–501 (2017).
3. Wijaya, Y. P. *et al.* Comparative study on two-step concentrated acid hydrolysis for the extraction of sugars from lignocellulosic biomass. *Bioresour. Technol.* **164**, 221–231 (2014).
4. Kanchanalai, P., Temani, G., Kawajiri, Y. & Realff, M. J. *Hydrolysis & crystallinity. BioResources* vol. 11 (2016).

5. Camacho, F., González-Tello, P., Jurado, E. & Robles, A. Microcrystalline-cellulose hydrolysis with concentrated sulphuric acid. *J. Chem. Technol. Biotechnol.* **67**, 350–356 (1996).

477a Effects of Sequential Pretreatment of Bamboo on Fermentable Sugar Yields for the Production of Biofuels and Bioproducts

Nneka Ekwe, Maksim Tyufekchiev, Ali Salifu, Klaus Schmidt-Rohr, Zhaoxi Zheng, Alex Maag, Geoffrey Tompsett, Wole Soboyejo, Michael T. Timko

Alternative feedstock for biofuel production based on non-edible plants and agricultural wastes should be explored to diffuse the food versus fuel conflict in the use of food crops such as corn and sugarcane for biofuel production. This is especially relevant in low- and middle-income countries where food security must be balanced with efforts at decarbonization. Utilizing plants such as bamboo that grow well on marginal lands that are not economically viable for cultivating food crops is crucial to diffusing this conflict. Unfortunately, bamboo has been studied far less as a renewable carbon source, unlike other feedstock such as corn stover and poplar. Thus, in this work, we studied bamboo endemic to West Africa (*Bambusa vulgaris*) as a feedstock for the production of renewable fermentable sugars. We investigated the effects of sequential pretreatment of bamboo prior to enzymatic hydrolysis as a way of maximizing the fermentable sugar yield to make it a cost-effective source of renewable carbon for the production of biofuels and bioproducts.

Initially, the bamboo was pulverized using a hammer mill, ground with a coffee grinder, and sieved to a particle size range of 212 – 425 μm. Then, the bamboo particles were ball milled for 30 minutes (BM30) and 60 minutes (BM60). Bamboo was also subjected to a sequential pretreatment where delignification pretreatment was carried out before ball milling for 60 minutes (DL-BM60), or delignification pretreatment was done after the ball milling step (BM60-DL). Delignification was accomplished using sodium chlorite and acetic acid. Afterward, we carried out enzymatic hydrolysis of the pretreated bamboo samples using Cellic CTec2 cellulase enzymes (20 FPU/g biomass) in 50 mM citrate buffer (pH 5.0) at 50°C for 72 hours and analyzed the hydrolysates using HPLC.

XRD analysis showed that ball milling pretreatment decreased the crystallinity of the bamboo (amorphization), with the BM60 pretreated samples being more amorphous. This translated to a recovery of 62% of the available carbohydrates in bamboo as fermentable sugars, primarily in the form of glucose and xylose, for BM60 versus a 40% sugar recovery for BM30 after enzymatic hydrolysis. Further analysis with solid-state 13C NMR revealed that ball milling for 60 minutes resulted in a near-complete cellulose amorphization, leaving only ~ 10% in the crystalline state. In contrast, the hemicellulose and lignin fractions underwent minor (<5%) transformations during ball milling. Thus, the higher sugar yield for BM60 was attributed to increased accessible surfaces for enzyme reaction coupled with the increased cellulose amorphization, which is a key indicator of cellulose reactivity.

This was followed by an evaluation of sequential pretreatment (DL-BM60 or BM60-DL) to address the two critical factors affecting the enzymatic saccharification of lignocellulosic biomass – cellulose crystallinity and lignin content – to maximize the fermentable sugar yield obtainable from bamboo. We observed through HPLC analyses of the hydrolysates that up to 80 ± 5% of sugar yields is obtainable when delignification was done after ball milling (amorphization), i.e., BM60-DL, with the sugar yields increasing to 97 ± 4 % when a reversed order of sequential pretreatment (DL-BM60) is implemented. Solid-state NMR analysis of delignified bamboo indicated a >95% lignin removal, resulting in a 16 percentage point increase in sugar yield (versus BM60). BM60-DL increased the sugar yield by only 2 percentage points.

XRD and solid-state NMR analysis results revealed that cellulose crystallinity increased after delignification and that DL-BM60 was more amorphous than BM60-DL. This meant that ball milling after delignification pretreatment (DL-BM60) was necessary to realize the combined effects of lignin removal and cellulose amorphization to achieve near-quantitative sugar yields (97%) from bamboo.

Finally, techno-economic analysis for bamboo conversion was done using a model developed by NREL for corn stover conversion to sugars as a starting point. The scale of the biorefinery was adjusted for handling a circular area with a 15 km radius centered on the refinery, with all other parameters held constant from their set points. The pretreatment used for corn stover analysis was substituted with the sequential pretreatment processes for bamboo. The scale was re-adjusted to account for differences in productivity between bamboo and corn stover. All other parameters were held at their fixed point to allow a direct comparison. Costs and labor were based on the U.S. market. The projected net present value (NPV) of the bamboo biorefinery was positive ($190 MM, U.S.), whereas that of a corn stover biorefinery was negative NPV (-$430 MM, U.S.) as a result of the higher growth rate of bamboo and the effectiveness of the sequential pretreatment.

In conclusion, this study revealed that ball milling bamboo for 60 minutes results in near complete amorphization of cellulose and increases the accessible surface for enzyme reactivity but does not achieve quantitative sugar yields after enzymatic hydrolysis, with almost 40% of unrecovered sugars left in bamboo. Therefore, sequential ball milling (amorphization) and delignification (lignin removal) is necessary to increase the sugar yields substantially, but the order of sequential pretreatment is crucial for maximizing the sugar yield. There is a clear preference for delignification followed by amorphization, i.e., DL-BM60, which should be factored into any technological deployment. Techno-economic analysis for bamboo conversion to sugars indicated that a bamboo biorefinery based on the preferred sequential pretreatment (DL-BM60) was economically viable with a projected positive NPV. These findings support further investment in bamboo cultivation in marginal lands and conversion to fermentable sugars as a renewable carbon source to decarbonize transportation fuels and chemical production without competing with food, especially in low- and middle-income countries.

477b Application of Three-Component Dihyroxybenzoic Acid-Based Deep Eutectic Solvent in a Sustainable Biorefinery Process

Jiae Ryu, Yunxuan Wang, Xianzhi Meng, Yang Tian, Linjing Jia, Aymerick Eudes, Kwang Ho Kim, Yunqiao Pu, Gyu Leem, Deepak Kumar, Arthur Ragauskas, Chang Geun Yoo

In this study, 3,4-dihydroxybenzoic acid (DHBA)-based three-component deep eutectic solvents (DESs) were used to pretreat transgenic sorghum bagasse with lignin rich in DHBA. DESs were formed with choline chloride (ChCl), DHBA, and a third component (water, ethanol, ethylene glycol, respectively). Without the addition of the third component, ChCl and DHBA formed DESs at molar ratios of 2:1 and 3:2. DESs formed with both molar ratios were evaluated for delignification and xylan removal of sorghum bagasse, and ChCl-DHBA formed at 3:2 resulted in better pretreatment performance. Recently, an increasing number of studies have reported the utilization of three-component DESs in biomass pretreatment. However, studies on three-component lignin-derived phenolic compound-based DESs are limited. To further increase the pretreatment performance, three-component DESs were synthesized based on ChCl-DHBA 3:2 DES and applied to pretreat both wild-type and transgenic sorghum bagasse. Chemical compositions, cellulose properties, and enzymatic digestibility of various sorghum bagasse samples before and after pretreatment were characterized. Cellulolytic enzyme lignin and fractionated lignin were isolated and structurally characterized to provide a comprehensive investigation of three-component DHBA-based DES. The results of this study will aid in the development of a sustainable biorefinery process using lignin-based DESs.

477c Comparative Study of Hydrothermal and Nades-Assisted Deconstruction of Transgenic Crops for Enhanced Recovery of Lipids and Sugars

Tirath Raj Sr., Vijay Singh

Modern technological progress, global climate concerns and the exhaustion of fossil-based feedstocks for energy and chemicals have driven the demand for viable alternatives to produce energy, chemical and materials. Thus, the development of efficient conversion technologies for second-generation biofuels has received greater interest from modern society to meet the upsurge in energy demand with bioeconomic developments. In view of this, newly developed energy crops, i.e., miscanthus, oilcane, and energy cane have been genetically engineered to simultaneously accumulate vegetative lipids and carbohydrates in their plant tissues [1, 2], thus can be considered as an alternative feedstock to cater for the enhanced biofuel yield by providing lipids along with cellulosic sugars for largescale biodiesel and bioethanol production [3-5]. However, the polymeric carbohydrate sugars are trapped in a recalcitrant lignin-carbohydrate matrix, which limits the efficient recovery of these components from transgenic crops for their downstream processing [6]. Therefore, the employment of an efficient pretreatment step is highly recommended to overcome the natural recalcitrant to facilitate the carbohydrate fractionation and their transformation into value-added chemicals, energy, and materials [7]. However, the conventional pretreatment methods involve the use of organic solvents, corrosive acids, and alkalis, which can be harmful to the environment and undermine the advocacy of a sustainable and circular economy. Besides, high energy, chemical input, production cost, inhibitor formation, low saccharification and fermentation yield are major key barriers to the development of efficient commercial biorefinery [7]. Thus, there is an immense need to develop and highly efficient fractionation of lipids and carbohydrates sugars for further processing into fuels and chemicals.

In view of this, hydrothermal and natural deep eutectic solvents (NADES) pretreatment potential were assessed for the deconstruction of genetically engineered bioenergy crops for their feasibility for high sugar and lipid recovery. Hydrothermal pretreatment is one of the mildest methods, which uses hot water at a relatively low temperature (up to 230 °C) and low pressure [8]. Hydrothermal pretreatment selectivity hydrolyses the hemicellulose fraction to xylose, arabinose and glucose and simultaneously generates a certain amount of degraded products such as furfural, HMF, formic acid etc., which can limit the enzymatic saccharification of lignocellulosic biomass [9]. On contrary, NADES is a combination of hydrogen bond donor (HBD) and hydrogen bond acceptor (HBA), with exceptional solvent characteristics, an alternative to conventional organic solvents and benefits with easy preparation, low toxicity, high biodegradability, and high fractionation efficiency. NADES can act as adjuvants for weakening the lignin-carbohydrate recalcitrance matrix at the desired temperature of 100 to 160 °C and pretreatment circumstances while enhancing lignin solubilization and could attribute to enhanced digestibility of pretreated biomass towards cellulases.

Thus, several choline chloride-based NADES were prepared using bio-derived precursors, i.e., lactic acid by varying the molar ratio of HBD and HBA (1:1, 1:2) at 60 to 80 °C and were employed for the pretreatment of transgenic oilcane bagasse, and miscanthus for the fractionation of lipids and carbohydrates. A comparative study between hydrothermal pretreatment and NADES-assessed pretreatment was conducted in terms of compositional analysis, toxins generation, chemical input, severity effect, lignin removal and overall sugar and lipid recovery from pretreated oilcane, and miscanthus biomass. The initial study showed that the suitable eutectic combination of NADES, i.e., choline chloride and lactic acid in a 1:2 molar ratio, prepared at 60-80 °C, could effectively solubilize >75% lignin, while enabling high biomass digestibility (>85%) during enzymatic saccharification, and enhance lipid recovery (>60%). In a separate experiment, hydrothermal pretreatment optimization was investigated with varying pretreatment temperatures (170-210 °C), with a residence time of 10-20 min and was compared with NADES pretreatment in terms of toxins release, glucan/ xylan recovery, lignin solubilization, crystallinity reduction, enzymatic accessibility, lipid recovery etc. Hydrothermal pretreatment of oilcane bagasse followed by enzymatic saccharification resulted in a 2-fold increase in total lipids with improved enzymatic digestibility of >75%. These findings lead to an economical an environmentally friendly method to process genetically engineered crops for selective fractionation of lipids and sugars, and would aid in developing a sustainable biorefinery platform for the co-production of biodiesel and bioethanol.

Keywords: Natural deep eutectic solvents; hydrothermal pretreatment, transgenic crop, lipids, sugar, extraction

References:

1. Parajuli, S., et al., *Towards oilcane: Engineering hyperaccumulation of triacylglycerol into sugarcane stems.* GCB Bioenergy, 2020. **12**(7): p. 476-490.
2. Kumar, D., et al., *Techno-economic feasibility analysis of engineered energycane-based biorefinery co-producing biodiesel and ethanol.* GCB Bioenergy, 2021. **13**(9): p. 1498-1514.
3. Jia, Y., S. Maitra, and V. Singh, *Chemical-free production of multiple high-value bioproducts from metabolically engineered transgenic sugarcane 'oilcane' bagasse and their recovery using nanofiltration.* Bioresource Technology, 2023. **371**: p. 128630.
4. Maitra, S., et al., *Bioprocessing, Recovery, and Mass Balance of Vegetative Lipids from Metabolically Engineered "Oilcane" Demonstrates Its Potential as an Alternative Feedstock for Drop-In Fuel Production.* ACS Sustainable Chemistry & Engineering, 2022. **10**(50): p. 16833-16844.
5. Cortés-Peña, Y.R., et al., *Economic and Environmental Sustainability of Vegetative Oil Extraction Strategies at Integrated Oilcane and Oil-Sorghum Biorefineries.* ACS Sustainable Chemistry & Engineering, 2022. **10**(42): p. 13980-13990.
6. Cortés-Peña, Y.R., et al., *Economic and Environmental Sustainability of Vegetative Oil Extraction Strategies at Integrated Oilcane and Oil-Sorghum Biorefineries.* ACS Sustainable Chemistry and Engineering, 2022. **10**(42): p. 13972-13979.
7. Raj, T., et al., *Lignocellulosic biomass as renewable feedstock for biodegradable and recyclable plastics production: A sustainable approach.* Renewable and Sustainable Energy Reviews, 2022. **158**: p. 112130.
8. Deshavath, N.N., et al., *A Chemical-Free Pretreatment for Biosynthesis of Bioethanol and Lipids from Lignocellulosic Biomass: An Industrially Relevant 2G Biorefinery Approach.* Fermentation, 2023. **9**(1).
9. Singh, R., et al., *Hydrothermal pretreatment for valorization of genetically engineered bioenergy crop for lipid and cellulosic sugar recovery.* Bioresource Technology, 2021. **341**.

477d Develop Deep Eutectic Solvent Systems for the Effective Fractionation of Lignin

Gloria Agyapong, Sampath Gunukula, Clayton Wheeler

Lignin is the second most abundant biopolymers in nature. The lignin content in softwoods is about 15 to 30wt% and 20 wt% in hardwoods. The lignin is currently underutilized because of a high heterogeneity in its structure. To utilize the great potential of this material fully, there is a need to develop an effective method that is economically and environmentally sustainable for producing lignin fractions with a more homogenous structure. In this work deep eutectic solvent systems with a lower viscosity will be developed, and its effectiveness in fractionating lignin will be studied. These novel deep eutectic solvents are characterized by measuring their melting point depression and the activity coefficients of each component in the system. Also, these solvents are further be characterized using the FT-IR technique. To understand the relationship between the solvent polarity of deep eutectic solvent systems and the chemical structure of the resulting lignin fractions, the collected lignin fractions and both the organosolv/native lignin are characterized using the NMR technique. The initial results indicated that both Organosolv and native lignin can be fractionated using the deep eutectic solvents with a low viscosity in two or more fractionation process steps. The solvent polarity of deep eutectic solvent is found to affect the physical properties of produced lignin fractions.

477e Fractionation of Cowpea Shells (*Vigna unguiculata*) Using Natural Deep Eutectic Solvent (Lactic acid-Proline)

Chioma M Onyelucheya, Joseph T. Nwabanne, Michael Daramola, Samuel Iwarere, Okechukwu Onyelucheya

For sustainability in biomass conversion, it is highly recommended to adopt pretreatment methods that do not threaten the environment. Natural deep eutectic solvents are made from substances found in plant cells Choi et al.(2011) and are environmentally friendly. They have been studied as a chemical route for biomass pretreatment because of their high solubilisation power and tunability (Abbott et al., 2011). These solvents are also easily produced by combining hydrogen bond donors and bond acceptors commonly found in plants in specific molar ratios.

This study applied a natural deep eutectic solvent that combined lactic acid and proline to fractionate cowpea shells. Cowpea (*Vigna unguiculata)* is one of the major sources of plant protein and generates a lot of waste from the deshelling process. According to FAO (2021), 8.9 million tonnes of cowpea were produced globally in 2020, of which Africa accounts for 97 %. Nigeria and the Republic of the Niger as the highest producers: 42 % and 31 %, respectively. A study by Jekayinfa et al. (2020) estimated that tonnes of cowpea shells are generated yearly in Nigeria and can serve as feedstock for biorefineries.

Factors such as temperature, time, and water dilution have been established as some of the factors that affect the use of NADES in biorefinery (New et al., 2019). The addition of water to solvents is done for several reasons, including reducing viscosity (Dai, Witkamp, Verpoorte, & Choi, 2015), to inhibit the solubility of cellulose. (Swatloski, Spear, Holbrey, & Rogers, 2002) and to reduce the volume of solvent used in a process (New et al., 2019). However, discrepancies in the research findings of different authors (Kumar, Parikh, & Pravakar, 2016; Li, Hou, Lin, Zhang, & Fu, 2018) on the effect of these factors necessitate this study.

Therefore, this work aims to apply NADES as a solvent for fractionating cowpea shells. The main objectives are to analyse the interactions between factors and to determine the optimal operating conditions using a central composite design of the response surface methodology. To the best of the authors knowledge, the effect of the interaction between temperature, time and water dilution on the fractionation of cowpea shells has not been explored. This study will provide more insight into applying NADES to pretreat cowpea shells.

Materials and Methods

The cowpea shells (*Vigna unguiculata* shells*)* were obtained from farms in Tunga in Chanchaga Local Government Area, Minna, Niger State, Nigeria and preserved in sample bags. Lactic acid (90 %) was purchased from Thermo Scientific (USA). Proline (99 %), Sulphuric acid (98 %), D- (+)-glucose and D- (+) -xylose were purchased from Sigma Aldrich Company Ltd, Germany.

The substrate was washed with deionised water and sundried for two days before milling with a Retsch PM 100 Planetary ball mill (Haan, Germany) and sieved with a 425 μm laboratory test before storage.

Lactic acid and proline were combined in a molar ratio of 4: 1 (Francisco, Van Den Bruinhorst, & Kroon, 2012) using the protocol described by Dai et al. (2013). The density, conductivity, and viscosity of the prepared NADES were analysed using the Attension Sigma force tensiometer (Sigma 700, Biolin Scientific USA), Jenco instruments (Model: 3020M) and RV-Viscometer (Model: NDJ-8S; W&J Instrument Co., Ltd. China), respectively.

The substrate was combined with the solvent at a 5: 100 (w/w), as determined by Kumar et al. (2016), in a capped Schott bottle and placed in a high-temperature bath circulator (Daihan Scientific, Korea). At the end of the experiment, the solid was recovered from the slurry based on the procedure reported by Procentese et al. (2015) and Li et al. (2018). The recovered solid residue was dried at 35 °C and analysed using the National Renewable Energy Laboratory (NREL) method (Sluiter et al., 2011) for its carbohydrate composition (glucan and xylan) and lignin composition. The upper and lower limits for the central composite design (CCD) were chosen based on the preliminary experimental results carried out. Temperature range (130 -150) °C, time (4 -6) h and percentage water dilution of solvent (1.5 – 3.5) % (w/w). The results were analysed using Design Expert software 13. All experiments were carried out in duplicates.

Results

The viscosity, conductivity, and density of the lactic acid-Proline solvent (4:1) prepared were recorded as 638.75 ± 0.35 mPa.s, 195.5 ± 0.28 μS/cm, and 1.26 ± 0.04 g/ml, respectively. The chemical composition of the cowpea shell was determined to comprise glucan (21.32 ± 0.39 %), xylan (21.46 ± 0.29 %), acid-insoluble lignin (28.40 ± 0.01 %), acid-soluble lignin (8.63 ± 0.02 %), extractives (2.10 ± 0.14 %), moisture (5.96 ± 0.65 %), Ash (1.05 ± 0.14 %). The central composite design was applied with design Expert 13 to obtain the quadratic model equation relating independent variables to the percentage responses measured (glucan, xylan and lignin). The statistical tests (ANOVA) were performed on the quadratic models to assess their accuracy based on the probability value (P) > 0.05, R-square (determination coefficient), adjusted R-square and predicted R-square. Percentage glucan yield in pretreated cowpea shell is significantly affected by the interactions between temperature and water dilution, time and water dilution. The interaction between temperature and water dilution significantly affects the xylan percentage composition of the pretreated sample. Lignin content percentage showed that interaction between temperature and time, temperature and water dilution, and time and water dilution were also statistically significant. These results imply that water dilution is key in lactic-proline solvent biomass pretreatment application. The process was optimised by applying the numerical method to obtain a solution satisfying the three quadratic equations by maximising glucan and xylan and minimising lignin. The optimal condition was 38.4 % glucan, 24.8 % xylan and 27.06 % total lignin at temperature: 146.4 °C, time: 6 h and water dilution at 1.5 % w/w. The models were validated by experimenting at optimal conditions. From the experimental results obtained, % glucan: 38.48 ± 0.21, % xylan: 24.66 ± 1.46 and % total lignin: 27.27 ± 1.24 shows that the model and the experimental values are in good agreement.

Understanding the effect of variation of factors on the ability of Lactic-Proline, natural deep eutectic on the fractionation of cowpea shells plays a crucial role in its industrial application. This study shows that NADES can be applied for the pretreatment of cowpea shells.

References

Abbott, A. P., Harris, R. C., Ryder, K. S., D'Agostino, C., Gladden, L. F., & Mantle, M. D. (2011). Glycerol eutectics as sustainable solvent systems. *Green Chemistry, 13*(1), 82-90.

Choi, Y. H., Van Spronsen, J., Dai, Y., Verberne, M., Hollmann, F., Arends, I. W. C. E., . . . Verpoorte, R. (2011). Are Natural Deep Eutectic Solvents the Missing Link in Understanding Cellular Metabolism and Physiology? *Plant Physiology, 156*(4), 1701-1705. doi:10.1104/pp.111.178426

Dai, Y., Van Spronsen, J., Witkamp, G.-J., Verpoorte, R., & Choi, Y. H. (2013). Natural deep eutectic solvents as new potential media for green technology. *Analytica Chimica Acta, 766*, 61-68. doi:10.1016/j.aca.2012.12.019

Dai, Y., Witkamp, G.-J., Verpoorte, R., & Choi, Y. H. (2015). Tailoring properties of natural deep eutectic solvents with water to facilitate their applications. *Food Chemistry, 187*, 14-19. doi:10.1016/j.foodchem.2015.03.123

FAO. (2021). Crops and livestock products. Retrieved from https://www.fao.org/faostat/en/#data/QCL

Francisco, M., Van Den Bruinhorst, A., & Kroon, M. C. (2012). New natural and renewable low transition temperature mixtures (LTTMs): screening as solvents for lignocellulosic biomass processing. *Green Chemistry, 14*(8), 2153-2157.

Jekayinfa, S. O., Orisaleye, J. I., & Pecenka, R. (2020). An Assessment of Potential Resources for Biomass Energy in Nigeria. *Resources, 9*(8), 92. doi:10.3390/resources9080092

Kumar, A. K., Parikh, B. S., & Pravakar, M. (2016). Natural deep eutectic solvent mediated pretreatment of rice straw: bioanalytical characterization of lignin extract and enzymatic hydrolysis of pretreated biomass residue. *Environmental Science and Pollution Research, 23*(10), 9265-9275. doi:10.1007/s11356-015-4780-4

Li, A.-L., Hou, X.-D., Lin, K.-P., Zhang, X., & Fu, M.-H. (2018). Rice straw pretreatment using deep eutectic solvents with different constituents molar ratios: Biomass fractionation, polysaccharides enzymatic digestion and solvent reuse. *Journal of Bioscience and Bioengineering, 126*(3), 346-354. doi:10.1016/j.jbiosc.2018.03.011

New, E. K., Wu, T. Y., Lee, C. B. T. L., Poon, Z. Y., Loow, Y.-L., Foo, L. Y. W., . . . Daud, N. N. N. (2019). Potential use of pure and diluted choline chloride-based deep eutectic solvent in delignification of oil palm fronds. *Process Safety and Environmental Protection, 123*, 190-198.

Procentese, A., Johnson, E., Orr, V., Campanile, A. G., Wood, J. A., Marzocchella, A., & Rehmann, L. (2015). Deep eutectic solvent pretreatment and subsequent saccharification of corncob. *Bioresource technology, 192*, 31-36.

Sluiter, A., Hames, B., Ruiz, R., Scarlata, C., Sluiter, J., Templeton, D., & Nrel, D. C. (2011). Determination of structural carbohydrates and lignin in biomass determination of structural carbohydrates and lignin in biomass. *National Renewable Energy Laboratory USA*, 1-15.

Swatloski, R. P., Spear, S. K., Holbrey, J. D., & Rogers, R. D. (2002). Dissolution of cellose with ionic liquids. *Journal of the American chemical society, 124*(18), 4974-4975.

477f Application of Biomass-Derived Solvent Cyrene in the Pretreatment of Lignocellulosic Biomass

Yunxuan Wang, Xianzhi Meng, Yunqiao Pu, Arthur Ragauskas

Dihydrolevoglucosenone (Cyrene) is a biomass-derived solvent that can be produced via the catalytic pyrolysis of cellulose followed by hydrogenation. It became industrially available recently and is a promising alternative to various toxic polar aprotic solvents in materials chemistry, organic synthesis, and other research fields. To exploit the potential of this novel green solvent and achieve a sustainable biorefinery process, herein, a biomass pretreatment process using Cyrene is developed. The fractionation of both hardwood and softwood under Cyrene-based organosolv systems was investigated. Results show that Cyrene effectively removed lignin and hemicellulose from the biomass at a mild condition (120 °C), leading to enhanced fermentable sugar yields after enzymatic hydrolysis. Lignin was recovered after pretreatment and structurally characterized to access its valorization potential and understand the interactions between Cyrene and biomass. Additionally, to better understand this novel solvent in biorefinery processes, investigation of the interactions between various components in the Cyrene-based solvent system was conducted, providing insight into future utilizations.

477g Optimization of Woody Biomass Conversion Via Amine-Based Solvent Pretreatment

Md Tahmid Islam, Venkataramana Pidatala, Hemant Choudhary, Roland Kalb, Blake Simmons, Alberto Rodriguez, John M. Gladden

The development of scalable and cost-efficient biomass pretreatment methods to enhance the conversion of biomass has been a challenging problem over the past decades. The purpose of this work is to optimize the conversion of woody biomass feedstocks (walnut, almond) to fermentable sugars when subjected to a process using distillable amine-based solvents, as well as demonstrating high solvent recovery and characterizing the resulting biomass fractions. Therefore, pretreatment experiments were performed at a bench-scale using an almond and walnut mix (80/20, w/w) with butylamine, ethanolamine,
ethylenediamine, triethylamine (boiling points; 78, 170, 117, and 89 °C, respectively), and amine-water mixtures at different weight ratios (10-75 wt%). The biomass-solvent ratio was maintained at a fixed value and the pretreatment was performed in a pressure tube reactor for a specified reaction time. The pretreated biomass slurry was processed in three pathways: (1) washed with water, (2) distilled (3) distilled and washed with water. Each of the fractions along with raw untreated biomass were characterized for glucan/xylan composition, CHN content, and sugar release efficacy. The glucan and xylan compositions suggested that the glucan/xylan content were improved by ~18-36 % with decreasing water percentage. Results indicate that pathway-1 showed >90 % sugar release, about 41-43 % higher than pathways-2 or 3. Elemental analysis via CHN analyzer showed that a reduction in C/N ratio follows their boiling point order after distillation. Washing of distilled biomass increases the C/N ratio by 19%, 384%, and 232% for butylamine, ethanolamine, and ethylenediamine, respectively, suggesting washing could hardly remove any butylamine as chemical bonding might be present. In addition, adding water in the amine-based solvent showed higher solid recovery, implying lower sugar release than standalone amine-based solvent. Ongoing solvent recycling experiments suggested promising recovery of these amine-based solvents after lignocellulosic biomass pretreatment.

Keywords: Biomass; Pretreatment; Amine-based solvent, Enzymatic hydrolysis; Sugar;

601a Cellulose Silylation in Flow with Enhanced Temperature and Pressure

Yue Yuan, Stylianos Andreou, Jong K. Keum, Sabine Neumayer, Rigoberto Advincula

Silylation is the introduction of a silyl group in substitution for active hydrogen in molecules. Because it reduces the polarity of the material surfaces and considering the abundant hydroxyl group (-OH) in cellulose structure, silylation has been widely utilized to produce modified cellulosic materials include but not limited to cotton fabrics, microcrystalline cellulose, bacterial cellulose and nanocellulose. Modified cellulose with higher hydrophobicity showed improved reinforcement performance. The silylated cellulose also showed reduced inter- and intramolecular hydrogen bonding, which alters the rheological behavior of materials during processing. However, the modification yields usually are not ideal and studies reported that the silylation often only happens at the substrate surface. Alternatively, flow chemistry offers well-controlled environment and the possibility of achieving high temperature and pressure in a very small reaction zone. This benefits the contact of reactants in the solution or suspension and uniformity of the reaction condition, compared to reactions in beakers or flasks with continuous stirring.

In this study, we conducted silylation reaction on cellulosic particles derived from different resources using amino silanes in a flow-chemistry set up. Silanes with different chemical structures, include 3-(2-aminoethylamino) propyltrimethoxysilane (AEAPTMS) and (3-Aminopropyl) trimethoxysilane (APTMS) were used as the reagent. We tested the flow reaction design with two scenarios: 1) passing reagent and cellulosic particles to the designated temperature region with defined volume, with or without increased pressure; and 2) fix the cellulosic particles in a region with designated temperature and pass reagent through a defined volume. The temperature was increased to above 100 °C, which is usually the limit of silylation temperature in literature. The surface chemistry of modified cellulosic substrates was characterized by Fourier Transform Infrared Spectroscopy (FTIR) and the nitrogen content was quantified via X-ray photoelectron spectroscopy (XPS). X-Ray Diffraction (XRD) was used to indicate changes of crystal structure and overall crystallinity of cellulosic substrates. Atomic Force Microscopy (AFM) was used to evaluate the morphology and mechanical properties of the unmodified and modified samples when nanocellulose were used as the substrates. The results suggested that the flow chemistry facilitate the silylation in bulk thus yielding a higher degree of substitution. In addition, the flow-reactor reduces the required time to achieve the substitution compared to reaction in batch. Therefore, flow chemistry offers the potential of large-scale modification of cellulosic materials, as well as providing novel molecular architectures in functional material innovation.

601b Impact of Sulfate Ester Content of Cellulose Nanocrystal on the Crystal Orientation and Its Applicability in Film Packaging

Ananya Ghosh, Zhihua Jiang

The modern world thrives towards green materials because synthetic materials deal with depleting fossil fuel resources while contributing to global warming, harming aquatic life, creating landfill conditions, and increasing energy consumption. In context to this, materials produced from bio resources could be an essential alternative to the synthetic ones. However, despite various advances in the use of biomaterials, there is a fundamental lack of knowledge about the quantitative relationships between the structural characteristics of biomaterials and their functional properties.

Cellulose nanocrystal (CNC) is a biomaterial produced from cellulose which is abundantly present in the earth. The application of such a material in the packaging industry could be advantageous when it is modulated in a way to achieve desirable properties. Typically, self-assembled films of CNC give rise to iridescence which reduces the transparency of the film. Further, having higher structural strength, CNC films become highly brittle in nature which limits their applicability. The overall goal of this project is to overcome these limitations of CNC films producing transparent and flexible films for packaging applications.

We are preparing CNC samples with varied sulfate ester and counter ion contents to systematically evaluate the fundamentals of how initial surface charge and counter ion content affect CNC nanoparticles interactions and self-assembly, and the structural, optical, mechanical, and barrier properties of CNC films. Additionally, we will be optimizing the use of plasticizing agents to improve the flexibility of CNC films for packaging applications.

601c Development of a Rechargeable Antimicrobial Textile Utilizing Radical Click Chemistry and Reactive Dyeing Techniques

Joseph Milter, Jason Stallings Jr., Thomas Glover, Terrence Ravine, Kevin West

Due to the moisture and heat provided by proximity to the human body, textiles serve as ideal habitats for microbes. Therefore, fabrics with the ability to self-disinfect have promising medical and military applications along with numerous commercial niches. Of the currently studied antibacterial agents, n-halamines are of particular interest due to their rechargeable nature and contact-killing efficacy. These biocides are created by the preferential uptake of cationic chlorine, usually sourced from hypochlorite/hypochlorous acid solutions, by certain amines, amides, and imides. When incorporated into materials, these compounds imbue a powerful contact-killing effect via the slow release of said chlorine. While this action does result in the re-conversion of halamine to amine, the process is amiably reversed by exposing the depleted compound to a similar chlorine source.

In developing a process for rending cotton with the properties of n-halamines, the most direct approach is covalent modification of the cellulose backbone. This strategy has been long-explored and is the basis of the reactive dyeing process which is used to impart color to textiles. Taking motifs from reactive dye chemistry, we designed a compound which employs cyanuric chloride as a binding agent. Cysteine serves as the link between this binder and the intended n-halamine which is built off 5,5-dimethylhydantoin. Our group had previously produced this molecule via an in-situ method, building directly off the textile surface, to investigate the retention of antibacterial effect after chemical grafting. Following success with the aforementioned approach, the subsequent goal was to create the agent as a standalone compound which can be applied to fabric in a single reaction. The devised synthesis strategy results in a chemical analogous to a dichlorotriazine (DCT) dye with suitable water solubility and reactivity for industrial textile manufacturing.

Said compound's structure is validated by FT-IR and NMR. Herein, the treated fabric is explored via FT-IR surface characterization to validate synthesis and ensure attachment. The effective chlorine loading of the treated fabric is determined through iodometry. This technique is also used to determine cycle rechargeability and storage stability. A novel qualitative antibacterial assay was designed to track contact-killing efficacy. In addition, a minimum inhibitory concentration study is performed to ratify the biocidal nature of all reagents used.

601e Janus Quaternarized-Hemicellulose-Modified PVDF Membrane for Desalination of Oily Seawater in Membrane Distillation

Rongrong Liu, Qun Li, Zhaohui Tong

Membrane distillation (MD) is an emerging thermally driven separation technology to recover pure water from seawater to alleviate severe water shortage problems. It is particularly attractive for its moderate operation temperatures (50-90 °C), which can utilize renewable or waste energy resources such as solar energy and geothermal energy, which contribute to reducing carbon emissions. However, conventional hydrophobic membranes are susceptible to membrane fouling and pore wetting when dealing with lightly polluted seawater (e.g., oil or other organic pollutants) due to hydrophobic-hydrophobic attractions. To address these challenges, asymmetric hydrophilic modification of hydrophobic membranes can effectively impede membrane failure by forming a hydration layer on the surface in the MD process. Hemicelluloses is a promising natural polymer candidate for hydrophilic modification in terms of the abundant hydrophilic hydroxyl groups, sustainability, and low cost. In this study, we used the hemicellulose that was extracted from the pretreatment wastewater of hardwood pulping process. Then, a Janus hydrophobic-hydrophilic dual-surface was constructed via a covalently anchoring hydrophilic layer with quaternarized hemicellulose, 2, 3-epoxypropyl trimethyl ammonium chloride (EPTMAC) modifying hemicellulose, through PVA and GA cross-linking on poly (vinylidene fluoride) (PVDF) membrane surface. The electrostatic interactions between negatively charged PVDF membranes and cationic hemicellulose can improve the grafting effects on the membrane surface and allow the formation of a stable hydration layer on the membrane surface. The quaternarized hemicellulose layer can provide double functions. First, the electrostatic interaction between quaternary ammonium salts and water molecules provides underwater oleophobicity and serves as an energetic barrier to prevent the adsorption of surfactants and oil droplets. Second, the positive charge on the membrane surface repels the mineral ions in seawater by electrostatic repulsion, improving the antifouling properties of the membrane and the penetration of pure water molecules. In comparison with the pristine PVDF membrane, the Janus membrane exhibited remarkably stable water vapor flux during long-term treatments of Tween® 80-stabilized oil-in-seawater emulsion in the bench-scale direct contact membrane distillation (DCMD) experiments. This study provides not only a high-value application of hemicellulose derived from the pulping process but also a sustainable fouling-resistant membrane for high-efficiency polluted salt water distillation.

601f Making Cellulose-Based Films Work in High Humidity

Tanner Hickman, Natalie Stingelin, Carson Meredith

Petroleum-based plastics are ubiquitous in society, but their limited recyclability and slow degradation present significant environmental concerns. In fact, most plastic waste is not recycled but instead discarded to landfills or accumulated uncontained in the environment where they persist for many years. Thus, solid waste accumulation, energy consumption, and CO_2 emissions associated with virgin plastic manufacturing have driven extensive research efforts to develop more sustainable alternative materials to traditional plastics. Cellulose is an obvious candidate to substitute or replace petroleum-based plastics due to its natural abundance, renewability, and biodegradability. Consequently, cellulose-based materials have been widely studied for applications such as packaging, composite materials, and printed electronics. However, the physicochemical properties of cellulose and its derivatives are sensitive to moisture and relative humidity, limiting their performance in many applications. Here, we discuss approaches to control swelling and water vapor transmission rates (WVTR) in cellulose-based films including crosslinking with renewable polyacid, coordination with soluble metal oxide hydrates, and polyelectrolyte complexation. We find that crosslinking with polyacid reduces swelling of cellulose-based films, and the WVTR of crosslinked films can be tuned over approximately four orders of magnitude by varying the crosslinker content. The WVTR of crosslinked films is one order of magnitude lower than that of poly(ethylene terephthalate), but increases when the RH is above 65% due to hydrophilic moieties of the polymer and crosslinker. Promisingly, the hydrophilicity of the crosslinker can be reduced by coordination with a metal oxide hydrate, preventing interaction with water. Similarly, polyelectrolyte complexes of cellulose and chitosan are less hydrophilic than the neat materials, demonstrated by their insolubility and high resistance to swelling. The approaches explored here to control moisture sensitivity in cellulose-based films are therefore promising for expanding the application space of cellulose as a sustainable alternative to petroleum-based plastics.

601g Biobased Carbon Nano-Onions and Their Applications in Nanoengineered Composites

Moham ED Abdur Razzaq, Xianglan Bai

Carbon nano-onions, consisting of quasi-spherical- and polyhedral-shaped graphitic layers close to one another, have many potential applications due to their unique structures. However, its high costs and the difficulty of scaled production hinder its wide applications. In this work, we developed an efficient and robust synthesis method to produce biobased carbon nano onions (BCNO) using lignin at low costs. Our BCNOs with tunable oxygenated functionalities and nanocrystal sizes also had excellent solubility in various solvents. As an example of its potential application, the BCNOs were used as fillers to develop property-enhanced PLA composites. It showed that adding 0.5% of BCNOs in PLA could increase tensile strength from 67 MPa to 92.8 MPa and tensile modulus from 2.5 GPa to 5.88 GPa. It was also found that adding 0.5% of BCNOs to a PLA composite with 10% soybean-based hyperbranched elastomer (SBCP) could increase tensile strength and modulus by 18.3% 179.5%, respectively, while also increasing tensile strain at break by 2225.3%, compared to pure PLA. Without BCNOs, PLA-SBCP composite increased the strain at break but reduced tensile strength and modulus. The addition to tensile properties, BCNO fillers could also improve the IZOD impact strength, flexural strength, and thermal stability of PLA composites. Our results show that BCNOs can be used as a low-cost filler to produce nanoengineered materials with superior performance.

601h Physical Activation of Microcrystalline Cellulose for High-Surface Area Carbon Materials

Katarina Odak, Kent J. Warren, Alan Weimer

Microcrystalline cellulose (MCC) is a promising precursor for additive manufacturing of active carbon due to its colloidal stability and ease of printing via extrusion-based methods. Carbonization of MCC prints can yield high-surface area carbon materials in complex shapes, offering a novel route to low cost, rapid manufacturing of detailed adsorbent structures with applications as filters, CO_2 capture, and reactive C templates. However, MCC carbonization is generally low yield (~10 wt%) and lengthy (~days-weeks).

In this study, we explore controlled carbonization in inert atmosphere to result in high-purity, high-surface area char from MCC in yields exceeding 20% and in less than 3 days. First, we use thermogravimetric analysis and in-line mass spectrometry (TGA-MS) to determine at what temperatures (< 300°C) is char formation favorable over tar formation. Next, we use TGA-MS and combustion analysis (for %C) to determine the effect of carbonization temperature on the time requirement for the low-temperature reaction. Then, we evaluate the efficacy of our low-temperature dwell after full carbonization (to 900°C) by characterization of the carbon content and specific surface area of the resultant char. Finally, we investigate the applicability of the carbonization schedule to a wide range of initial MCC sample size.

49a Biocarbon-Metal Sustainable Composites for Energy Storage

Ange Nzihou

Authors:

Theotime Béguerie[a], Amel Cydric Ghogia[a], Lina María Romero Millán[a], Majd El Saddik[a], Elsa Weiss[a], Claire E. White[b,c], Ange Nzihou[a, c*]

1. Université de Toulouse, IMT Mines Albi, RAPSODEE CNRS UMR 5302, Campus Jarlard, F.81013 Albi Cedex 09, France
2. Princeton University, Department of Civil and Environmental Engineering, Princeton, NJ 08544, USA
3. Princeton University, Andlinger Center for Energy and the Environment, Princeton, NJ 08544, USA

*Corresponding author: Email addresses: ange.nzihou@mines-albi.fr, ange.nzihou@princeton.edu

Abstract:

This presentation will discuss the synthesis, characterization, and utilization of a novel family of lignocellulosic biocarbon-based composites for electrochemical and heat storage. The composites are produced by the pyrolysis and carbonization of biomass inherently containing or doped with catalytic metal such as Iron and Calcium at lower temperature (<1400°C) than the standard carbonization temperature (>1800°C). In this process, the biocarbon is transformed from a randomly organized carbon to a graphite-like material (organized and structured carbon) showing a high graphene rate structure. The Carbon structure change in the composites has been investigated from nano to bulk scales in combining XRD, HRTEM and RAMAN. This has enabled to probe changes in biochar nanostructure and crystalline graphitic domains catalyzed by the inclusion of metal such as Fe and Ca [1-3]. The encapsulation of iron particles by graphitic phase (graphene sheets) and the carbon-metal bonding during the pyrolysis and carbonization were uncovered using XPS [3]. Key properties for energy storage applications such as the remanent magnetization, the coercive field, the specific heat, the thermal and electric conductivities, the thermal stability including the mechanical properties will be discussed.

References:

- Romero Millán, Lina; Ghogia, Amel; White, Claire; Nzihou, Ange. 'Iron Nanoparticles to Catalyze Graphitization of Cellulose For Energy Storage Applications'. ACS Applied Nano Materials 6, n°5 (February 2023): 3549–3559. https://doi.org/10.1021/acsanm.2c05312
- Beguerie, Théotime; Weiss-Hortala, Elsa; Nzihou, Ange. 'Calcium as an innovative and effective catalyst for the synthesis of graphene-like materials from cellulose.' Scientific Reports (December 2022): 21492. https://doi.org/10.1038/s41598-022-25943-3.
- Ghogia, Amel C.; Romero Millán, Lina M.; White, Claire E.; Nzihou; Ange. "Synthesis and Growth of Green graphene from Biochar Revealed by Magnetic Properties of Iron Catalyst." ChemSusChem (November 2022): 16. https://doi.org/1002/cssc.202201864.

49b High-Performance Silicon Active Materials from Biorenewable Resources

Srikanth Pilla

High-Performance Silicon Active Materials from Biorenewable Resources

Nancy Chen, Morteza Sabet, Craig M. Clemons, Apparao M. Rao, and *Srikanth Pilla

Ph.D. Candidate, Clemson University, Research Assistant Professor, Clemson University, Materials Research Engineer, USDA Forest Products Laboratory, Professor, Clemson University, and Professor, Clemson University

Presentation Keywords: lithium-ion batteries, anode, silicon nano-quill, sol-gel, cellulose nanocrystal

Abstract

The escalating global population and surging energy demands are straining our fossil fuel reserves and raising concerns about resource availability for future generations. As the transportation sector rapidly electrifies, rechargeable lithium-ion batteries (LIBs) are emerging as a viable alternative to conventional fuel-based technologies. The quest for higher energy density and longer cycling life continues to drive advancements in LIBs. Among the recent innovations in anode electrode materials, silicon (Si) is considered the most promising candidate to replace graphite, boasting a theoretical specific capacity of ~4200 mAh g^{-1}, over ten times greater than that of graphite anodes (~372 mAh g^{-1}). Despite these advancements, persistent challenges obstruct the commercialization of anodes based on pristine Si active material. Si-based electrodes are prone to rapid degradation due to the substantial volume change of Si particles (approximately 400%) during lithium insertion and extraction. This repeated volume change results in the pulverization of Si material, ultimately causing decreased cycling stability due to the loss of contact with the current collector. To overcome this obstacle, porous and hollow nanostructures have been employed to provide sufficient void space to accommodate the volume change during electrochemical cycling. However, with the demands for material cost-reduction from industry, Si structures' strategic engineering must be cost-effective for commercial viability. To this end, we developed a low-cost process in which we leverage the 1-dimensional (1D) bio-renewable templates to synthesize a 3D porous Si architecture called silicon nano-quills (SiNQs). We innovated a two-step, cost-effective process that yields SiNQs with a porous morphology and hollow interior structure. First, in a scalable sol-gel process, silica gel particles were prepared using low-cost chemicals. A unique mesoporous morphology was engineered using surfactant-modified cellulose nanocrystals as a sacrificial template. The templates were removed via thermal treatment to form silica nano-quills (SilicaNQs), which possess a 3D bulk structure comprised of hollow quill-like arms and a high degree of porosity. In the second step, we employed a low-temperature magnesiothermic reduction method to convert SilicaNQs into SiNQs with a relatively large surface area.

The electrochemical cycling performance of SiNQs was determined using half-cell testing. A water-based slurry was prepared using a combination of SiNQ and graphite as the active materials. The slurry with 73 wt% MCMB graphite, 15 wt% SiNQ, 2 wt% carbon black, and 10 wt% LiPAA binder was cast onto the commercial copper foil. After room-temperature drying, the electrode was calendered and vacuum dried. We also prepared identical electrodes using commercial Si nanoparticles (100 nm, solid spherical). The 2032-type coin half cells were assembled for battery testing using SiNQ-graphite and Si-graphite electrodes (with an active mass loading of 3 mg cm^{-2}). The coin cells underwent a formation cycle at a current rate of 0.05C followed by continuous cycling at a current rate of 0.1C (90 mA g^{-1}) over the potential range of 0.005 – 1.5 V at room temperature. The SiNQ-graphite anode offered an initial reversible capacity of 587 mAh g^{-1} and a superior capacity retention of 70% after 200 cycles. In comparison, the commercial Si-graphite battery exhibited an initial capacity of 375 mAh g^{-1} and capacity retention of 41% after 200 cycles. The SiNQs possess a BET surface area of 399 m^2 g^{-1} and a total pore volume of 0.64 cm^3 g^{-1}. The superior performance of SiNQs is due to their unique morphology that offers high surface area and porosity for effective diffusion of lithium ions and their electrochemical interactions with NQs, leading to a higher reversible capacity. Moreover, the porous architecture of SiNQs can effectively mitigate the volume change during lithiation and delithiation, thus ensuring excellent cycling performance.

49c From Waste to Treasure: Multiple-Functional Materials from Bio-Based Waste

Zhaohui Tong

Our society is facing long-term sustainability challenges such as greenhouse gas emissions, intensive energy use, resource depletion, food scarcity, water and soil contamination. To tackle these concerns and challenges, our team has been dedicated to synthesizing value-added functional biomaterials from renewable resources, especially bio-based waste from agricultural and forestry fields. The adaption of chemical functionalization and self-assembling methodologies to renewable starting materials for functional materials is very challenging due to bio-based feedstocks' macromolecular structures, heterogeneous properties, poor solubility, and the disturbance of impurities. In this talk, we will focus on how we explore new synthesis and functionalization methods/ reaction systems to attack these technical barriers. We will use some examples to describe a series of recently fabricated functional biomaterials for applications in the areas of environmental mitigation, food safety, and human health. For example, a glycerol-based intelligent and sensoring food packaging film for a universal, accurate, easy-to-use, and real-time food spoilage monitoring system. Besides, smart materials with controllable release functions have been synthesized from low-cost agricultural residues to improve nutrient (N, P, and micronutrients) utilization efficiency in agriculture and reduce environmental pollution to soil and water. Ultimately, we would like to utilize a waste-treat-waste approach to fabricate value-added functional materials to achieve a high-efficiency circular bioeconomy.

49d Circular Economy Driven Sustainable Composites through Waste Valorisation for Single-Use Plastic Alternatives

Manjusri Misra

Plastic waste is a global pollution problem that appears daily on news headlines, with microplastics even showing up in Mount Everest's snow. "New Plastic Economy" by Ellen MacArthur's Foundation targets "100% reusable, recyclable or compostable plastic packaging by 2025". The application of plastics can be found in many sectors and products including, but not limited to, food packaging, transportation, medical, housing, and other high-end products. Packaging is the major sector for plastic use; ~50% of which is in single-use products like grocery bags, takeout containers, straws, six pack rings and plastic cutlery. Plastic wastes account for ~300 million tons annually, which is equivalent to the weight of the entire human population (United Nations). Thus, immediate and vigorous action is required in reducing plastic wastes. There remains a huge challenge in alternative solutions for three categories of packaging – multi-layer packaging, mixed-material (plastic, metal, and paper) and mono-material packaging that are impractical to recycle. Biodegradable plastics show leading trends in sustainable packaging. The global bioplastic market is growing at a CAGR of ~12% and several biodegradable plastics are being commercialized through cost competitive blends and biocomposites. The biodegradable plastic market is expected to grow to a ~$5 billion market in 2025. The end-of-life solution for biodegradable plastics should grow and should be expanded through government and industry leaders as well as policy makers. Special interest is growing around home compostable and soil biodegradable products over industrial composting. Disruptive technology in creating high barrier biodegradable packaging can help in reducing food wastes. This presentation will highlight challenges and opportunities towards a zero-waste plastics economy in supporting circular economy. The combination of circular economy with bioeconomy can lead to advanced sustainability.

590a Comparative Techno-Economic and Lifecycle Assessment of Electrochemical Processes for Lignin Valorization

Zahra Ebrahimpourboura, Manish Mosalpuri, Cheng Yang, Marcus Foston, Elijah Thimsen, Mark Mba Wright

Lignin is an abundant renewable resource that can processed into biofuels, chemicals, and other high-value-added products. Lignin is commonly combusted in biorefineries to produce heat and power. Higher value applications could improve the profitability and sustainable of second generation biorefineries. This study evaluates the commercial potential of electrodeless electrochemical (EC) conversion as a strategy for lignin valorization to produce commodity and specialty chemicals.

In this study, a techno-economic (TEA) and lifecycle (LCA) evaluation of a lignin valorization process was carried out in order to determine its feasibility and to demonstrate lignin's potential as a value-added product. We developed a chemical process model in BioSTEAM based on the National Renewable Energy Laboratory (NREL) biorefinery design. A portion of the lignin stream was diverted to the EC unit for chemical production.

Experimental data for this analysis is based on phthalimide-N-oxyl (PINO) type catalysts. The performance of the PINO-type catalysts mediated electrochemical oxidation was evaluated to determine the yield and selectivity. Through integrating chronoamperometry, all of the charges flowing through the electrochemical cell were determined (i-t curve). Accordingly, Faraday's effectiveness was calculated.

We determined that a 2,000 metric tonne per day (tpd) biorefinery could produce between 4.4 and 37 tpd of chemicals by processing 50% of process lignin. The primary chemical by mass is syringaldehyde (16.6 tpd), and vanillin is the most valuable by revenue ($206,000/day). The bulk chemical minimum selling price ranges between $1.15 and $7.10 per kg. We estimated the maximum EC capital investment to range $5 and $50 million for chemical prices of $1 to $10/kg. These results indicate that a EC unit could generate additional revenues to second generation biorefineries. Future studies will evaluate the impacts of product composition on the commercial potential of this technology.

590b Enabling Commercially Viable Green Carbon Fiber By Tuning Lignin Chemistry

Yixin Luo, Moham ED Abdur Razzaq, Xianglan Bai

Carbon fiber is popular in applications where the ratio of tensile strength to weight is important, such as the aerospace, automobile, wind turbine, construction, infrastructure, and sporting goods industries. For example, lightweight vehicles partly constructed with carbon fiber improve fuel efficiency, reducing greenhouse gas emissions from the transportation sector. Despite increasing demand, carbon fiber is mostly applied to higher-end products due to its high costs attributed to its precursor costs. In this context, low-cost carbon fibers produced using polyacrylonitrile-alternative precursors are attractive in various industrial applications. According to DOE, low-cost carbon fibers with a tensile strength of 1.72 GPa and a modulus of 172 GPa could be widely used in automobile industries for bodies and parts without compromising performance. Lignin is a promising precursor for low-cost green carbon fiber due to its abundant availability at low costs and high fixed carbon content. Unfortunately, its poor mechanical properties hinder the market entry of lignin-based carbon fiber.

This talk will introduce our approaches for producing high-quality lignin-based carbon fibers. The poor mechanical properties of lignin-based carbon fibers are mainly attributed to the irregular and highly branched molecular structure of lignin lacking in orientation. To address this intrinsic defect of lignin, we previously synthesized melt-spinnable arylate thermoplastics with linearity from lignin bio-oil by applying a controlled radical polymerization method. Using the new thermoplastic polymer as a precursor, we achieved carbon fibers with an average tensile strength of 1.7 GPa and tensile modulus of 182 GPa. In a recent study, we innovated carbon fiber production by developing a thermo-mechanically controlled and lignin-tailored method. By coupling the thermochemistry and mechanochemistry of lignin, we manipulated chemical reactions and altered the microstructure evolution. As a result, lignin-based carbon fibers with unprecedented tensile properties were obtained at surprisingly low temperatures. Three carbon fibers produced using raw lignins without premodifications or additives had tensile strength and modulus of 2.45 GPa and 236 GPa, 2.11 GPa and 215 GPa, 2.2GPa and 225GPa, respectively. Our findings suggest promising opportunities to produce commercially viable low-cost green carbon fibers from broad lignins, advancing lignin valorization.

590c Fabrication and Property Testing of Carbon Foams Using 100% Kraft Lignin

Qiangu Yan, Zhiyong Cai

Lignin is one of the major components of lignocellulosic biomass and the most important natural phenolic polymer. It is a wood by-product composed of 60-65wt% carbon with an annual production over 70 million tons worldwide from the paper and pulping industries [1]. Much research has been done on the processes involving lignin thermal decomposition to sustainable/renewable carbon-based materials like active carbons, carbon fibers, templated carbon, and graphene [2], but less work has been done on carbon foam production from lignin. Mansmann and Winter [3] developed a process to produce carbon foam from an aqueous lignin solution. However, only ligninsulfonates were used as the lignin precursors, polyethylene oxide and acrylic acid-acrylamide were added as the copolymer, and tremendous energy was consumed to vaporize water from the ligninsulfonate solution in the heating process. Spradling and Amie [4] described a process to produce carbon foam from lignosulfonate. The lignin resource was limited to lignosulfonates in both previous works [3, 4] while Kraft lignin, which comprises 95% of the lignin produced annually, was not used for carbon foam production [5]. Currently, carbon foams are mainly produced by the process of blowing the carbon precursors, which are usually first decomposed in a closed vessel at high temperature, under high pressure [6]. This accounts for the high cost of the product since high temperature and high-pressure facilities are required for production. Furthermore, the product size and properties are limited by the facility [7]. Furthermore, catalysts, foaming/blowing agents, surfactants, and/or crosslinking agents may be used for the foam preparation [8], this further increases the cost of the product.

This work reports a simple manufacturing process to produce carbon foams in an open vessel using Kraft lignin as the only carbon precursor; no solvents, catalysts, blowing agents, surfactants, and crosslinking copolymers are needed in this developed method. The process includes pressing the carbon foam precursors into a lignin block in a mold at room temperature, then the lignin block is heated to create a lignin foam block at a temperature range of 400-600 °C, and the resultant lignin carbon foam (LCF) is finally obtained by carbonization at 1100 °C. The properties obtained and structural examination showed a density of 0.18 – 0.68 g/cm^3; porosity of 69.3 – 91.5%; open-cell rate of 95%-98%; pore size ranges from 10 to 1000 μm; pore wall thickness is distributed between 10 and 100 μm; compressive strength of 7.03 ± 1.25 MPa – 30.16 ± 2.41 MPa; thermal conductivities of LCF samples increased with increasing foam bulk density; the bulk electrical conductivities of LCF samples also increased with increasing bulk density. Lignin carbon foam blocks also showed high levels of both fire and termite resistance. Fire resistance testing showed no damage occurred when the LCF sample was heated with an acetylene/air flame over 1000 °C for more than 3 minutes. Termite testing showed that termites ignored the lignin carbon foam blocks, but readily consumed the pine feeder block as a food source.

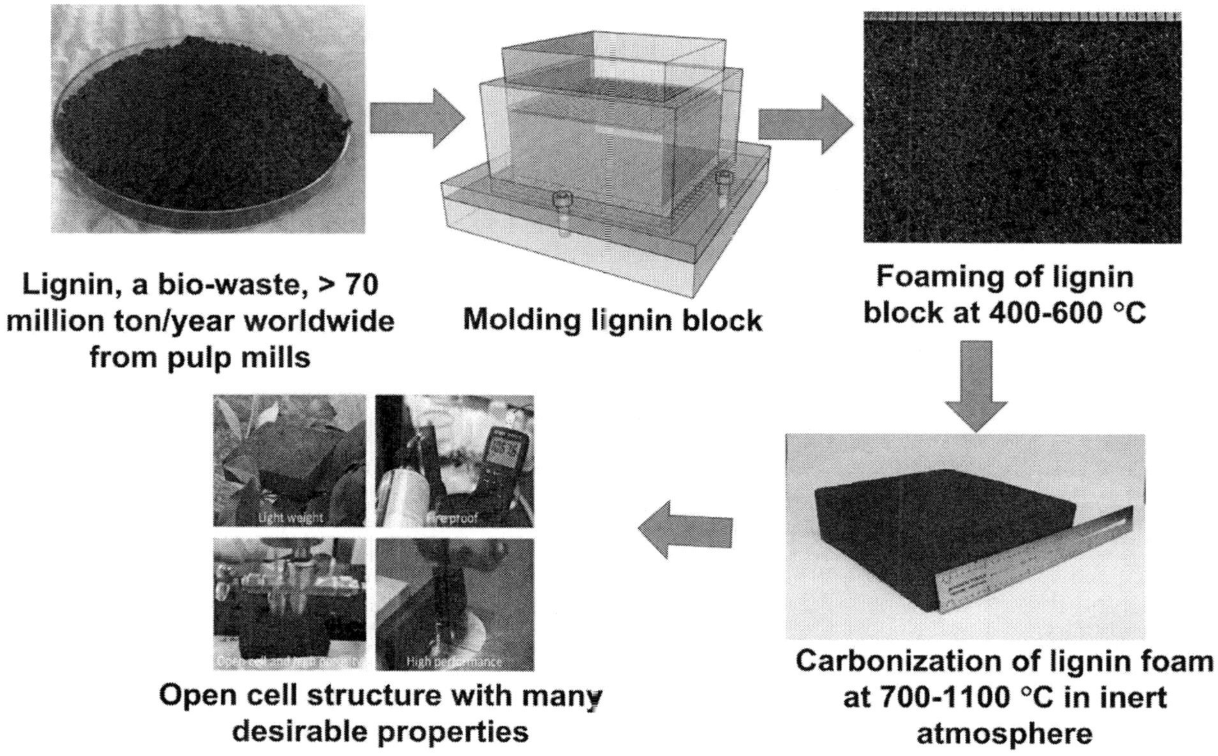
Lignin, a bio-waste, > 70 million ton/year worldwide from pulp mills
Molding lignin block
Foaming of lignin block at 400-600 °C
Carbonization of lignin foam at 700-1100 °C in inert atmosphere
Open cell structure with many desirable properties

590d Economic and Environmental Impact of Recovering and Upgrading Lignin Via the ALPHA Process on an Ethanol Biorefinery

Robert Handler, Daniel Kulas, Graham W. Tindall, Mark C. Thies, David Shonnard

Despite its prevalence as one of the world's most abundant organic polymers, a small minority of lignin is recovered for nonfuel uses. Many challenges face lignin recovery and valorization, including achieving low ash and metal contents and a tight molecular weight distribution. In this study, we investigate economics and environmental impacts of one emerging method to valorize lignin: the Aqueous Lignin Purification with Hot Agents (ALPHA) process. The ALPHA process has been proven capable of fractionating and purifying of raw, bulk lignins recovered from cellulosic ethanol biorefineries or Kraft pulp mills. The ALPHA process exploits the novel liquid-liquid equilibrium that exists between lignin and hot, one-phase solutions of aqueous renewable solvents to simultaneously purify and fractionate raw bulk lignin. Through sequential ALPHA reactors with controlled liquid-liquid equilibrium conditions, distinct fractions of lignin can be produced with different molecular weight profiles, while also controlling the distribution of sugars and ash within the separation system. These tailored fractions can be sold as high-quality feedstock for producing polyurethane foam, activated carbon, carbon fiber.

Our analysis considers the recovery of lignin from two lignin-rich waste streams from an ethanol biorefinery: black liquor and lignin cake. The black liquor is subjected to three sequential ALPHA stages with varying ethanol/water solvent concentrations, which produces lignin suitable as a polyurethane foam precursor, activated carbon precursor, and carbon fiber precursor. Two different lignin cake scenarios are analyzed: selling the lignin cake directly as an activated carbon feedstock with no processing and subjecting the lignin cake to one ALPHA stage which produces a refined lignin for both polyurethane foam and activated carbon processes. Using inputs generated at the lab scale, an ASPEN Plus simulation was developed to model a 130,000 dry metric tons lignin/yr plant that valorizes the waste lignin from an ethanol biorefinery. Economic results generated using a 30-year discounted cash flow table show a net present value greater than $100 million for the simulated scenarios. Environmental impacts were calculated using life cycle assessment methodology and the software SimaPro. The fractionated lignin products have GHG emissions and cumulative energy demand equivalent to or lower than the fossil precursor equivalents. Coupling the economic and environmental results show that using the ALPHA process is a promising emerging technology to valorize waste lignin.

590e Designing Ion-Conducting Materials for High-Temperature Fuel Cell Applications Using Low-Cost, Biorenewable Lignin

Rajesh Keloth, Shudipto K. Dishari

High-temperature proton exchange membrane fuel cells (HT-PEMFCs) have many benefits, such as faster electrode kinetics, high tolerance to fuel impurities, and simplified water management. Acid-doped polybenzimidazole (PBI) membranes are promising candidates for HT-PEMFCs due to high proton conductivity at elevated temperatures and good chemical and thermal stability. However, PEMFCs operated at high temperature suffer from leaching of doped acid due to weak acid-base interaction between PBI and phosphoric acid (PA) affecting the membrane performance. Several approaches have been adopted to improve PA retention within the PBI membrane matrices. In these efforts, quaternary ammonium cation-containing species appear to be ideal candidates for their favorable interactions with biphosphate anions (of PA) preventing the leaching of PA. This talk will show our efforts in designing cationic version of lignin to improve the PA capture and holding within the PBI membranes. Lignin is the second most naturally abundant polymer (next to cellulose). Lignin is also a by-product of pulp and paper industries, cellulosic biorefineries, and agricultural farms. However, most cases, this biomass remains unvalorized and thus wasted, whereas the three-dimensional, hyperbranched structure of lignin can favor the formation of unique ion conduction pathways. By functionalizing neutral lignin with cationic quaternary ammonium groups and incorporating cationic lignin within PBI membrane matrices, we were able to improve the PA retention by the composite PBI membranes which in turn improved the proton conductivity of the PBI membranes even at high temperature. Additionally, the cycling experiments showed that the cationic lignin-PBI-based composite membranes retained the high protonic conductivity at high temperature (140 °C) over an extended period of time (240 h). The stable and high conductivity thus suggested the effectiveness of cationic lignin in boosting the ion transport performance of high temperature PEMFCs in a green manner.

590f Advancing Lignin Valorization: Elucidating Thermodynamic and Kinetic Properties through an Integrated Density Functional Theory and Kinetic Monte Carlo Approach

Chi-Ho Lee, Juhyeon Kim, Joseph Kwon

Lignin, a complex aromatic polymer found in plant cell walls, is a highly attractive feedstock for producing renewable chemicals and fuels due to its abundant availability and significant potential for valorization[1]. As the second most abundant polymer on Earth after cellulose, lignin plays an important role in plant growth, development, and defense against pathogens[2]. Nevertheless, the complex structure of lignin currently poses a substantial challenge for the development of sustainable materials and processes, as its resistance to degradation complicates its utilization. Therefore, understanding the de/repolymerization behaviors of lignin is essential for devising more efficient and sustainable valorization processes, which could potentially reduce our dependence on fossil fuels[3].

In this study, we employed a multifaceted approach to investigate the de/repolymerization of lignin by integrating density functional theory (DFT) method and kinetic Monte Carlo (kMC) simulations, which allowed us to consider various factors affecting the de/repolymerization behavior. Specifically, DFT calculations enabled us to predict structural changes, thermodynamic properties, and reaction pathways of lignin de/repolymerization at the atomic scale[4-5]. However, DFT alone is insufficient for modeling the kinetics of lignin systems (i.e., lignin monomer yield and chain length distribution) at the microscopic level. This is where kMC simulations prove invaluable, as they facilitate a deeper exploration of the kinetics of lignin de/repolymerization reactions by simulating the time evolution of complex systems[6].

To accurately interpret the results of kMC simulations, it is necessary to integrate the DFT method, as it provides critical information on lignin de/repolymerization, such as activation energy barriers and their variations according to lignin size and changing structural conformations. This information can then guide the interpretation of kMC simulations, offering insights into the time-dependent concentration of lignin fragments and the distribution of lignin sizes.

By employing integrated DFT and kMC simulations, we aimed to obtain a more accurate interpretation of lignin de/repolymerization reactions under various conditions, including temperature and lignin size. Specifically, we examined the temperature dependency of activation energy barriers required for de/polymerization reactions and how the size and structural conformation of lignin can influence this dependence. We believe that our comprehensive, multifaceted approach provides valuable insights into the thermodynamic and kinetic properties of lignin systems, which will ultimately aid in the development of valuable products (i.e., carbon fiber, phenolic compounds, multifunctional hydrocarbon and syngas) more effectively.

References

[1] Zeng, Y., Himmel, M. E., and Ding, S-Y. Visualizing chemical functionality in plant cell walls. *Biotechnol Biofuels*, **2017**, 10, 263.

[2] Satteler, S. E, and Funnell-Harris, D. L. Modifying lignin to improve bioenergy feedstocks: strengthening the barrier against pathogens? *Front. Plant Sci*, **2013**, 4.

[3] Iram, A, Berenjian, A, and Demirci, A. A Review on the Utilization of Lignin as a Fermentation Substrate to Produce Lignin-Modifying Enzymes and Other Value-Added Products. *Molecules.* **2021,** 26(10), 2960.

[4] Mu, X., Han, Z., Liu, C., and Zhang, D., Mechanistic Insights into Formaldehyde-Blocked Lignin Condensation: A DFT Study. *J. Phys. Chem. C,* **2019,** 123, 8640-8648.

[5] Azad, T., Auad, M. L, Elder, T., and Adamczyk, A. J., Toward Native Hardwood Lignin Pyrolysis: Insights into Reaction Energetics from Density Functional Theory. *Energy Fuels,* **2023,** 37, 401-423.

[6] Eswaran, S. C. D., Subramaniam. S., Sanyal, U., Rallo, R, and Zhang, X., Molecular structural dataset of lignin macromolecule elucidating experimental structural compositions. *Nature,* **2022,** 9, 647.

590g Incorporation of Unmodified Technical Kraft Lignin Particles in Anticorrosive Epoxy Novolac Coatings

Tejasvi Laxminarayan, Alessio Truncali, Narayanan Rajagopalan, Claus Erik Weinell, Mats Johansson, Søren Kiil

To protect steel structures from corrosion in harsh environments, high-performance anticorrosive coatings are commonly used. However, due to an increasing demand for environmentally friendly alternatives, there is a need to incorporate renewable raw materials in coating formulations.

The present study aims at replacing pigments and fillers with technical Kraft lignin particles in epoxy novolac coatings. Furthermore, the influence of Kraft lignin particles on the anticorrosive properties of the coating were evaluated and compared with an epoxy novolac coating containing iron oxide pigment (IO-EN), a commercial coating and a neat (particle-free) epoxy novolac coating (neat EN), using salt spray exposure, rust creep assessment (ISO 12944-9), and chemical resistance test. Methods, such as scanning electron microscopy (SEM), size exclusion chromatography (SEC), differential scanning calorimetry (DSC), pull-off adhesion, and König pendulum hardness, were used to study the thermal and mechanical properties of lignin-based epoxy novolac coatings.

Compared to the IO-EN coating and commercial coating, the lignin-based EN (L-EN) coating showed about 6% and 8% lower rust creep after 70 days of salt spray exposure. However, no surface defects or chemical degradation were observed on any of the coatings.

The L-EN coating showed an excellent adhesion strength (23 MPa) and impact resistance (0.49 N), exceeding the values for both commercial (17 MPa and 0.41 N) and lignin-based diglycidyl ether bisphenol F (L-DGEBF) coatings (13 MPa and 0.07 N). Furthermore, the addition of lignin particles did not influence the chemical resistance, hardness, or glass transition temperature of the epoxy novolac coatings.

This study highlights the potential physical and chemical interactions between lignin particles and epoxy novolac, and elaborates the orientation of lignin particles within the coating matrix. Finally, Kraft lignin particles, without solvent fractionation and/or chemical modification, can be incorporated into heavy-duty coatings as a bio-based alternative to pigments and fillers, thereby presenting an opportunity to reduce the use of fossil-based ingredients in anticorrosive coatings.

590h Improvement of Thermal and Mechanical Properties of Lignin Copolymers and Synthetic Polymer/Lignin Copolymer Blends

David Chem, Samantha Glidewell, Keisha Walters

Lignin is arguably one of the most abundant bio-based polymers found in nature, and it can be extracted from the lignocellulosic part of the plant. Lignin is made up of 15% to 35% of the biomass, depending on the type of plant. It has a cell wall structure that is responsible for the rigidity and strength of the plant. With its abundance, inexpensiveness, and biodegradability, it has been used in blending with synthetic polymers to obtain new potential polymeric materials that have comparable chemical and mechanical properties to synthetic homopolymers. However, lignin and synthetic polymer blending display a non-uniform distribution, low miscibility, and poor properties in blends. Therefore, lignin must be chemically modified or used in conjunction with a compatibilizer when blending with synthetic polymers. The objective of this research is to synthesize copolymers from lignin and synthetic polymer – resulting in a potentially biodegradable polymer containing improved chemical and mechanical properties that can also be used in blending with synthetic homopolymers. Activators Generated by Electron Transfer Atomic Radical Polymerization (AGET ATRP) is the method used in this work due to its high molecular weight control, better oxygen insensitivity, less toxicity, and high industrial applicability. The three chosen monomers in this work are methyl methacrylate (MMA), methacrylic acid (MAA), and styrene (S) because of the hydrophobicity and similarity in their structures. The lignin copolymers synthesized from this work are tested to determine the chemical and thermal properties through FTIR, NMR, DSC, and TGA analysis. Their mechanical and surface morphology properties are investigated through tensile testing, optical microscopy, and SEM analysis. Lignin copolymers show an enhancement in thermal properties, such as increased thermal stability and higher glass transition temperature compared to native lignin. Moreover, due to the lignin backbone, lignin copolymers also show biodegradability. When used in a blend with synthetic homopolymers, they display an improvement in miscibility and mechanical properties compared to based lignin and synthetic polymer blending. This work will promote more use of lignin in the renewable plastic industry, and it will also benefit the agricultural economy.

590i Kinetic Modeling and Optimization of a Bio-Adhesive Production Process from Lignin

Poulomi Das, Debangsu Bhattacharyya, Changle Jiang, Jianli Hu

Due to the gradual depletion of fossil fuel sources along with the issues of greenhouse gas emission and resulting environmental impact, interest in the development of sustainable technologies has rapidly increased over last few decades. The global wood-adhesive industry heavily depends on the petroleum-derived commercial adhesives like phenol–formaldehyde (PF), urea–formaldehyde (UF) and melamine–urea–formaldehyde (MUF)[1]. These formaldehyde-based chemicals cause hazards to both human health and environment through emission of toxic volatile organic compounds[1,2]. Hence, this project focuses on manufacturing of bio-based adhesives using renewable materials such as soy-protein and lignin.

Lignin, one of the main constituents of lignocellulosic biomass, is considered to be a potential raw material for bio-adhesives due to the presence of both alcohol- and phenol-type hydroxyl groups in its complex molecular structure[3]. In this work, kraft lignin is used as the feedstock for the BCD process. Each year, more than 50 Mt kraft lignin (KL) is produced as a waste in the paper and pulp industry, and almost 98% of it is burned as a fuel causing high emission of greenhouse gases[3–6]. Production of the bio-adhesive from KL consists of two steps. In the first step, the lignin is depolymerized to smaller oligomer units. These oligomer fragments are added to the soy protein isolate in the second step to produce the bio-based adhesive. The degradation of lignin can be achieved by various thermochemical, electrochemical, or biological treatments[4] in presence or absence of catalysts. The thermochemical method includes hydrothermal base catalyzed depolymerization (BCD) process. BCD of lignin can generate various multicomponent and multiphase products[3,7]. In spite of several existing kinetic models[7,8] for depolymerization of lignin under hydrothermal conditions either in presence or absence of acid catalysts, only a handful of papers have developed a kinetic model for the hydrothermal BCD of lignin[5]. These existing kinetic models have been developed in terms of lumps like oligomers, liquid oil and SOC, gaseous products etc., and the properties of all compounds of a lump are considered to be same[3,5]. In this work, reaction rate models are developed for most key components and by including the concentration of the base catalyst in the kinetic model. Kinetic parameters are optimally estimated by using the experimental data. Models of the separation section are developed for separating the oligomers from other undesired products. These oligomers are then cross-linked with soy protein isolate (SPI) to produce the bio-adhesive. A kinetic model of the reactor for producing bio-adhesive is proposed and rate parameters are estimated by using the experimental data. The lab-scale reactor models are scaled to the commercial-scale and models of additional balance of plant are developed for mass and heat integration and recycle of catalysts. A economic model of the process is developed and a mathematical optimization problem is set up to optimize the design parameters and operating conditions of the reactors for maximizing the net present value.

References:

1. Gonçalves, D., Bordado, J. M., Marques, A. C. & Dos Santos, R. G. Non-formaldehyde, bio-based adhesives for use in wood-based panel manufacturing industry—a review. *Polymers (Basel).* **13**, (2021).
2. Thuraisingam, J., Mishra, P., Gupta, A., Soubam, T. & Piah, B. M. Novel natural rubber latex/lignin-based bio-adhesive: synthesis and its application on medium density fiber-board.

Iran. Polym. J. (English Ed. **28**, 283–290 (2019).
3. Ročnik, T., Likozar, B., Jasiukaitytė-Grojzdek, E. & Grilc, M. Catalytic lignin valorisation by depolymerisation, hydrogenation, demethylation and hydrodeoxygenation: Mechanism, chemical reaction kinetics and transport phenomena. *Chem. Eng. J.* **448**, (2022).
4. Agarwal, A., Rana, M. & Park, J. H. Advancement in technologies for the depolymerization of lignin. *Fuel Process. Technol.* **181**, 115–132 (2018).
5. Bernhardt, J. J., Rößiger, B., Hahn, T. & Pufky-Heinrich, D. Kinetic modeling of the continuous hydrothermal base catalyzed depolymerization of pine wood based kraft lignin in pilot scale. *Ind. Crops Prod.* **159**, (2021).
6. Mahmood, N., Yuan, Z., Schmidt, J. & Xu, C. Production of polyols via direct hydrolysis of kraft lignin: Effect of process parameters. *Bioresour. Technol.* **139**, 13–20 (2013).
7. Yong, T. L. K. & Matsumura, Y. Reaction kinetics of the lignin conversion in supercritical water. *Ind. Eng. Chem. Res.* **51**, 11975–11988 (2012).
8. Zhang, B., Huang, H. J. & Ramaswamy, S. Reaction kinetics of the hydrothermal treatment of lignin. *Appl. Biochem. Biotechnol.* **147**, 119–131 (2008).

240a Metal Ion Modification Enabled Hydrophobic and Dimensionally Stable Wood

R.M. Oshani Nayanathara, Xuefeng Zhang

Wood is a hygroscopic material that responds to the moisture changes of the surrounding environment through swelling and shrinking, making it dimensionally unstable. This property is attributed to hydrophilic groups of wood's cellulose, hemicellulose, and lignin components. In this study, we introduce a facile metal-ion-modification (MIM) approach to enhance the dimensional stability of wood. The MIM process involved swelling the wood samples with aqueous metal ion solutions and drying. The high valent metal cations, such as Fe^{3+}, Al^{3+}, and Zr^{4+}, interacted with the hydrophilic groups (e.g., OH, COOH) present in the wood fibers, limiting their access to water and moisture and thereby enhancing the wood's hydrophobicity and dimensional stability. Three wood species including southern yellow pine, poplar, and red oak were treated by MIM. The dimensional stability of wood samples was evaluated by measuring their full swelling ratios. The MIM treatment significantly decreased swelling in both the tangential and radial directions. For example, after Fe^{3+} treatment, the tangential swelling ratio of southern yellow pine decreased from 6% to 4%. Moreover, all the Fe^{3+}-treated wood samples (southern yellow pine, poplar, and red oak) exhibited tangential anti-swelling efficiencies of 39.83%-57.14% and radial anti-swelling efficiencies of 34.74%-48.33%. Overall, our findings highlight the effectiveness of the MIM method in improving the dimensional stability of wood and providing a new route for wood property enhancement.

240b Impact of Methoxy Group Component of Lignin-Derived Deep Eutectic Solvents on the Biomass Pretreatment Performance

Jiae Ryu, Jialong Lin, Yunxuan Wang, Gyu Leem, Kwang Ho Kim, Arthur Ragauskas, Chang Geun Yoo

Deep eutectic solvents (DESs) can be readily synthesized with relatively low costs compared to ionic liquids (ILs). We have investigated lignin-derived DESs as a green technology for biodegradability and biocompatibility along with the valorization of lignin. The chemical structure of hydrogen bonding donor (HBD) and hydrogen bonding acceptor (HBA) are crucial to determine the performance of DESs. In this study, lignin-derived HBDs which are methoxy-substituted phenol structures were applied to synthesize DESs, and hydrogen bonding formation and physical and thermal properties of DESs were studied. In addition, the impacts of HBD structure on each DES performance with biomass conversion were investigated. DESs including phenol, guaiacol, and syringol as a model compound were evaluated by differential scanning calorimetry (DSC), thermal gravimetric analysis (TGA), rheometer, and nuclear magnetic resonance (NMR). The performances of DESs on sugar maple were evaluated by the chemical compositions and sugar conversion of cellulose-rich fraction after DESs pretreatment. The recovered lignin was characterized with the GPC and 2D HSQC.

240c Effects of Lignin Properties on the Performance of Lignin-Based Polyurethane Foam for Eco-Friendly Thermal Insulation

Soyeon Jeong, Mairui Zhang, Charles M. Cai, Xianzhi Meng, Paul Crovella, Paul Crovella, Arthur Ragauskas, Chang Geun Yoo

Polyurethane (PU) is widely used as thermal insulation in various applications due to its low thermal conductivity, density, and cost-effectiveness. However, their production heavily depends on petroleum-based products like polyethylene glycol (PEG), which has negative environmental impacts and poses potential health risks. Lignin, a bio-based resource, is a highly promising candidate for use as polyols in polyurethane production, owing to its high content of hydroxyl groups and aromatic structures that can be easily converted into polyols. Even though lignin has great potential to be used as a component for polyurethane foam production, its limited reactivity and poor dispersibility require further modification to improve its performance. In this study, low-molecular and high-molecular weight lignins were fractionated from bulk lignin extracted by co-solvent enhanced lignocellulosic fractionation (CELF) pretreatment from poplar, corn stover, and pine. The lignin-PU foams were produced with each lignin fraction, and their physical and thermal properties were evaluated to determine the optimal conditions with suitable properties as insulation materials.

240d Use of UV Irradiation on Photo Isomers to Enhance the Removal of Lignin from Woody Biomass Hydrolysate

Dipesh Karki, Sampath Gunukula, Clayton Wheeler, G. Peter Van Walsum

ABSTRACT

Problem:

Conversion of lignocellulose to bioproducts and biofuels is often initiated with a hydrolytic pretreatment process that helps to depolymerize the native polymers in biomass and produce a cellulose-rich pulp stream. Extracting dissolved components from this hydrolysate contributes to deriving value from biomass components other than cellulose.

The mildly hydrophobic components in the hydrolysate, such as dissolved lignin, and dehydration products, such as furfural or levulinic acid, are often recovered using liquid-liquid extraction. Recovery of the desired extracts from the solvent typically involves distillation or back extraction, both of which have high energy demands. In this investigation, we propose a novel method to enable the precipitation of desired solutes using photo-isomers in the extraction solvent. The azobenzene moiety is stable in its linear trans-isomer, but will isomerize when irradiated with UV light into the more polar/bulkier cis-isomer. The cis-isomer will likely interact differently with the extraction solvent, which could change the solubility of the lignin in the solvent. This investigation sought to determine if the presence of a photo-isomer in a solvent could result in a reversible UV light induced precipitation of lignin.

Methods:

Three solvents: hexanol, ethanol, or acetone, were combined with three different photo-isomers: azobenzene, diethylamino azobenzene, or diethoxy-azobenzene, at varying concentrations, to observe the precipitation behavior of lignin dissolved in the solvent-isomer mixture. The lignin used in these experiments was derived from the Organosolve process, which was one motivation for the selection of ethanol and acetone as potential solvents to be studied. Hexanol solvent was chosen as a representative of biphasic liquid liquid extraction of biomass hydrolysates.

Solubility limits of the three photo-isomers in different solvents were determined at different temperatures, as was the solubility of Organosolve lignin in each of the solvents. Experiments to test precipitation were carried out at concentrations at or below the solubility of the photo-isomers and the lignin.

The charging (isomerizing trans to cis) was done by exposure to 365 nm UV light, whereas dis-charging (re-isomerizing cis to trans) occurred through exposure to visible light. Experiments were carried out either in glass beakers that were irradiated with UV light on the upper surface of the solvent-isomer solution, or by using quartz beakers that could be irradiated through the vessel walls. This latter arrangement enabled sealing off the top of the beaker to minimize evaporation of the solvent. Precipitation of lignin under irradiation was observed visually and the amounts of precipitate were quantified by filtering the precipitated solution and measuring the mass of dried lignin.

Results:

Experiments with this system provided insights on solubility concentrations, precipitation, and charging/discharging time durations. Using the simple azobenzene molecule was found to be impractical for developing a precipitation process because the highly rapid rates of isomerization and de-isomerization made the handling too cumbersome. Hexanol has very low solubility of both the photo-isomers and the lignin, and thus was not suitable for extracting or precipitating lignin.

It was found that among the different combinations of solvent and photo-isomer, the acetone and diethoxy azobenzene solution was most efficient at precipitating dissolved lignin in response to UV light exposure. It was found that the UV-induced precipitation was stable for some time, and that the solubility of the system could be restored by shutting off the UV light and allowing the isomerization to reverse and increase the lignin solubility. The concentrations of diethoxy azobenzene in acetone were varied, and the amount of precipitation was recorded. It was observed that precipitation increases with an increased amount of photo-isomer up to a certain point, above which a "shading effect" comes into play, whereby a high concentrations of photo-isomer in one region of the solvent container appears to block the penetration of the UV light, and thus isomerization in the "shaded" regions of the vessel becomes limited.

Implications:

The experimental work validated our hypothesis that a switchable solvent system could be developed that makes use of UV light as the means of activating a switch in solvent activity.

Making use of a UV activated switchable solvent may enable new approaches to separations of different products in dilute solutions. Organosolve lignin in particular may be a suitable material to be separated from biorefinery process streams.

240e Application of VBA and Excel to Develop Ternary Diagram of Brine, Oil, Green Corrosion Inhibitor from Lignocellulosic Resource

Tianxing Cai

VBA (Visual Basic for Applications) and Excel are powerful tools that can be used for various applications in chemical engineering. They can be used in chemical engineering, especially for Data Analysis (Excel is commonly used in chemical engineering to analyze data from experiments and simulations. VBA can be used to automate data analysis and to create custom data analysis tools.) and Visualizations (Excel can be used to create visualizations of data and simulations. VBA can be used to create interactive visualizations.). In this study, VBA and Excel are used to develop a tool of ternary diagram for brine, oil, and green corrosion inhibitor from lignocellulosic resources. The necessary data for the ternary diagram are given by experiment measurement. This data will include the concentrations of the brine, oil, and corrosion inhibitor, as well as their respective densities and mass/volume in each liquid phase. The ternary diagram can be customized, by adding labels and adjusting the color scheme as necessary. The tool for ternary diagram of brine, oil, and green corrosion inhibitor can be created in this way. Green corrosion inhibitors are biodegradable and do not contain heavy metals or other toxic compounds. The physical properties such as concentration, solubility and multi-phase equilibrium has a great impact on the inhibition effect.

240f Fused Deposition Modeling (FDM) 3D Printing of Industrial Hemp Waste Polymer Composites

Anqi Ji, Nara Han, Shuyang Zhang, Samarthya Bhagia, Xianhui Zhao, Arthur Ragauskas, Deepak Kumar, Jeong Jae Wie, Chang Geun Yoo

Three-dimensional (3D) printing through fused deposition modeling (FDM) is a promising and material-efficient technology. Industrial hemp waste, the hemp waste after extracting CBD oil, is considered as non-utilized biomass and directly goes to landfill. Hence, this study aims to use industrial hemp waste as filler to enhance polymer properties for high value-added applications. However, the high extractives content of the hemp waste can cause difficulties in its extrusion and 3D printing due to nozzle clogging, insufficient mechanical strength, and other technical challenges.

In this study, the industrial hemp waste was treated with sodium hydroxide (NaOH) to enhance its performance as a filler by controlling its protein content for making 3D printing products with polylactic acid (PLA). The goal of this project is to study the feasibility of fabricating industrial hemp waste/PLA composites and also investigate the effect of filler content on the mechanical and thermal properties of the biocomposites. The results indicated that by removing the extractives, the compatibility of hemp waste with plastics would be improved, thus enhancing the interactions between hemp waste and polymer matrix, ultimately improving the properties of 3D printing composites.

240h On the Effects of Chaotropic Promoters on Enzyme Hydrolysis

Timothy Woodard, Michael T. Timko, Aidin Panahi

Lignocellulosic biomass, being an abundant source of non-edible carbon is an ideal feedstock replacement for petroleum, originating in the form of biomass such as grasses, wood, and agricultural waste. This type of biomass is mainly composed of cellulose, hemicellulose, and lignin while the general method for the conversion of biomass to biofuel precursors consists of a pretreatment step followed by enzymatic hydrolysis into simple sugars, that can later be fermented into bioethanol. The pretreatment leading up to the conversion step removes lignin from the feedstock, leaving cellulose and hemicellulose, following up with preparation of the cellulose to be in a more reactive amorphous structure to later be used to create biofuel precursors.

Electrolytes such as chaotropes from the Hofmeister series salts have been shown to increase the solubility of organic molecules in water solutions. These substances can disrupt the hydrogen bond network in liquid water, thereby increasing entropy. Previous attempts to utilize the benefits of hydrolyzing amorphous cellulose have been unconcluded by the spontaneous recrystallization of cellulose to a less reactive form, by inhibiting the recrystallization using a chaotropic promoter, the process can take advantage of the benefits from the enhanced cellulose reactivity. Recent studies have shown that implementing Hofmeister series salt promoted cellulose hydrolysis reactions resulted in an increase in glucose yields by 80%, making the bioethanol production more economically favorable.

In this study, the focus is on understanding the effects of the Hofmeister series salt by exploring the chaotropic salt guanidinium chloride's interaction with the enzyme used during hydrolysis with the use of cellobiose as the model reaction. Cellobiose is a water-soluble carbohydrate consisting of two glucose molecules that reproduces the glycosidic bonding of cellulose. As a water-soluble molecule, cellobioses reactivity is not affected by crystallinity or the effects of a water-solid interface.

The glucose yields at the different salt concentrations of guanidinium chloride for a reaction time span of 0 to 12 hours were studied and compared in which it was found that the glucose yields varied slightly (~10%) between the control enzyme hydrolysis sample with no salt added and the optimal sample promoted with 0.2 molar guanidinium chloride as shown in *Fig. 1.*

Figure 1. Glucose Yields from Cellobiose Enzymatic Hydrolysis

This study aims to shed light on the effects of Hofmeister series salts on the reaction by looking at cellobiose to see how the salt interacts with the enzyme reactivity. These results imply that other parameters like crystallinity may have an impact and more work is needed to be done to understand how the Hofmeister series salts are affecting the reactions. In doing so this process has the potential to unlock cost-effective biofuels at an industrial scale.

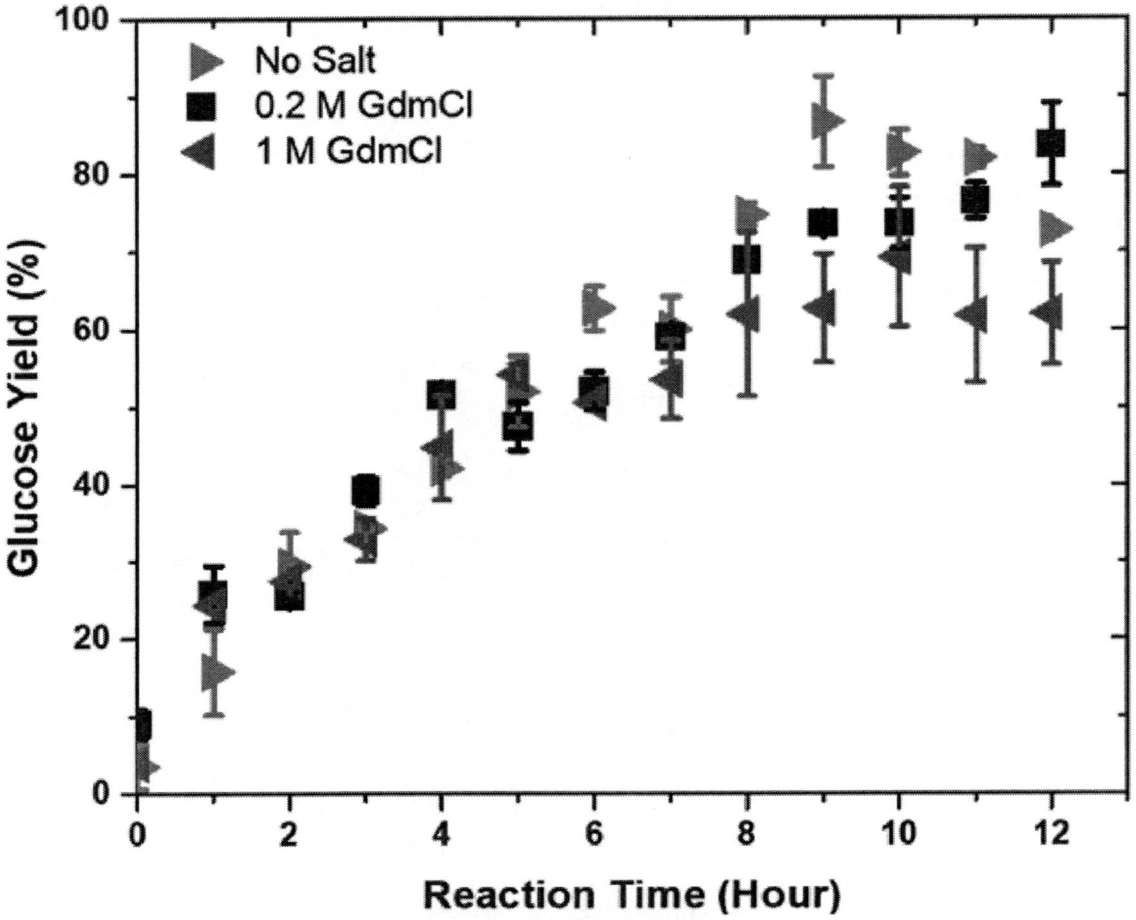

Glucose Yield (%)
Reaction Time (Hour)
No Salt
0.2 M GdmCl
1 M GdmCl
0
20
40
60
80
100
0
2
4
6
8
10
12

240i Mechano-Chemical Analyses of Plant Cell Walls at Nanoscale Using Atomic Force Microscopy Coupled with Infrared Spectroscopy

Huiyong Li, Ram Dixit, Natasha Bilkey, Anton Ievlev, Neus Domingo Marimon, Marcus Foston

Plant cell walls play a critical role in protecting against environmental stress, providing structural support, and determining the rate and direction of cell growth. The organization of plant cell wall components plays a role in determining the local mechanical properties of the cell wall. As a result, the natural processes of plant cell and tissue growth, development, and function rely on control of cell wall composition, arrangement, and ultrastructure to manipulate local mechanical properties. Investigating the role of local chemical composition on the local mechanical properties of the plant cell wall requires that methodologies probing the cell wall chemical and mechanical properties be at a length scale smaller than the size of the plant cell. In this study, methods of atomic force microscopy coupled with Infrared spectroscopy (AFM-IR) was developed to perform nanoscale multimodal imaging of the physical, chemical, and mechanical features of plant cell walls in the *Arabidopsis* stems and roots, and poplar tension wood. AFM-IR not only overcomes the diffraction resolution limit of other chemical imaging techniques like IR microspectroscopy, but also characterizes mechanical and chemical properties of the samples simultaneously at nanoscale. Using multivariate statistical analysis tools, we were able to quantify compositional variations and mechanical heterogeneity, and to correlate the chemical composition and local mechanical properties in *Arabidopsis* and poplar cell walls.

409a Modeling and Simulation of Drying of Paper and Board

Hua-jiang Huang, Shri Ramaswamy

Paper and board continue to be an important renewable bioproducts manufacturing sector in the US and the world. Conventional drying processes used in majority of the paper and board are based in conduction heat transfer from steam heated dryer cans and convective heat and mass transfer from air flowing over the surface of the paper in the pockets as it is being dried. The drying process, as in many process industries, is one of the major consumers of energy. Approaches to reducing energy consumption and possible decarbonization are of considerable interest. Fundamental understanding of the transport processes involved in conventional drying processes in porous materials like paper and board can help us develop ways to reduce energy consumption and decarbonization of the drying process. Here we present preliminary results from the modeling and simulation of the heat and mass transfer and associated liquid, gas and energy fluxes during the drying process. This will help in better understanding of the conventional processes and the development of novel drying processes.

409b Experimental Analysis of Drying of Paper and Board

Koushik Sampath, Leonard Reynolds, Hua-jiang Huang, Shri Ramaswamy

In conventional multi cylinder drying of paper and board both conductive drying from steam heated dryer cylinders and convective drying by flowing air over the surface in the pockets are used. Conductive drying from steam heated drying cylinders is a critical component in providing the necessary thermal energy to paper and board as they are dried. Steam temperature, internal and external resistances at the contacting surface are critical process parameters influencing the conductive drying process. An experimental setup was developed to study the conductive drying of paper and board. Paper sheet moisture, temperature as well as the temperature distribution within the heated platen and the instantaneous heat flux as the sheet is being dried were measured. The instantaneous heat flux, contact heat transfer coefficient and the drying rates were determined as drying proceeds. Experimental results as well as comparison to literature and commercial data are presented. The conductive heat transfer coefficients determined were compared to traditional correlations normally used in the modeling of paper drying. We are also developing a combined conductive and convective drying of paper and board and the potential inclusion of auxiliary energy components. Preliminary experimental results from the conduction and convection drying system will also be presented. This data will be useful in process development, intensification of manufacturing processes and modeling and simulation of paper drying processes.

409c The Investigation of Various Biomass Feedstocks for the Synthesis of 1 – 3 Layers Graphene

Haider Al-Rubaye, Paul Ani, Hasan Al-Abedi, Joseph Smith

Graphene is a two-dimensional form of crystalline carbon, either a single layer of carbon atoms forming a honeycomb (hexagonal) lattice or several coupled layers of this honeycomb structure. Graphene is a parent form of all graphitic structures of carbon: graphite, which is a three-dimensional crystal consisting of relatively weakly coupled graphene layers.

Few-layered graphene has been synthesized by numerous methods including mechanical cleavage, liquid-phase exfoliation, gas-phase synthesis, Hummers' method, unrolling of multi-walled carbon nanotubes (MWCNTs), chemical vapor deposition (CVD), epitaxial growth, an electrochemical reaction. However, these methods have some drawbacks, such as low yield ratio, high-energy consumption, the use of expensive substrates, as well as the difficulty of obtaining a high-quality product.

The LPE technique is based on the exfoliation of graphite intercalated compounds in an appropriate solvent. In this technique, there are three subsequent steps: (1) dispersion of feedstock in an acidic medium solution (distill water and hypochloric acid), (2) exfoliation, and (3) collecting. For the intercalation of the acid-based solution molecules, the Van der Waals forces between the carbon layers are overcome by the application of the external driving force, which is the ultrasound energy.

409d Transgenic Energycane As an Alternative Feedstock for Biodiesel Production: An Industrially Relevant Study

Shraddha Maitra, Ming-Hsun Cheng, Hui Liu, Dang Viet Cao, Baskaran Kannan, Stephen P. Long, John Shanklin, Fredy Altpeter, Vijay Singh

Lignocellulosic biomass has abundant renewable carbon that can be converted conveniently to biofuels such as bioethanol, biodiesel, and bio-jet fuel. Non-transgenic bioenergy crops have shown great potential for providing cellulosic sugars for bioethanol and high-value bioproduct production. To further increase the energy density of plants, high biomass bioenergy crops such as energycane, sugarcane, and sorghum, are being genetically modified to sequestrate carbon in the vegetative tissues and divert the carbon flux to synthesize and accumulate triacylglyceride (TAG) molecules (Luo et al., 2022; Parajuli et al., 2020; Vanhercke et al., 2019). These transgenic crops cater to both cellulosic sugars and vegetative lipids. Simulations based on lab-scale experiments with transgenic crop suggests that these transgenic crops have the potential to replace oilseeds as feedstocks for biodiesel production (Clark and Schwender, 2022; Huang et al., 2016; Maitra et al., 2022b). Recently, energycane has been successfully engineered to accumulate 30-fold higher TAG molecules and double the amount of total fatty acids in green leaves as compared to their wild-type variety. The hyperaccumulation of TAG molecules was achieved by overexpression of the lipogenic factors *DGAT*1-2 (*Diacylglycerol acyltransferase*1-2) and *OLE*1 (*Oleosin*1) along with the suppression of SDP1 (*SUGAR-DEPENDENT*1) and TGD1 (Luo et al., 2022). The amount of total fatty acid and TAG of the transgenic lines varied from 3.56 to 4.96 % and 0.11 to 1.52% of leaf dry weight, respectively (Luo et al., 2022).

However, since lipids in the vegetative tissues of transgenic energycane are present in lower amounts and are in a complex form (Maitra et al., 2021; Parajuli et al., 2020), appropriate and efficient bioprocessing technologies are needed to deconstruct the lignocellulosic structure and recover vegetative lipids without degenerating along with cellulosic sugars from the feedstock. Chemical-free hydrothermal pretreatment has been shown to maintain the lipid profile during processing and enrich biomass residues with vegetative lipids that can be recovered at the end of the process (Maitra et al., 2022b; Maitra et al., 2022a). The initial characterization of transgenic energycane and process optimization at the lab scale has been established (Luo et al., 2022; Maitra et al., 2022b; Maitra and Singh, 2021).

We are presenting the proof-of-concept of bioprocessing transgenic energycane at a pilot scale to demonstrate its potential commercialization as an alternative renewable feedstock for biodiesel production. The transgenic energycane plant was segregated into stems, green leaves, and brown leaves in the field during harvesting. Juice was extracted from stems. Pilot-scale continuous hydrothermal pretreatment reactor was used to pretreat bagasse at 50% solids loading at 190 °C for 10 minutes followed by disk milling. The same pretreatment conditions were used to pretreat green and brown leaves at the lab scale. Chemical-free pretreatment had no adverse effect on the percent recovery and composition of total fatty acids and TAG fatty acids. After pretreatment, a major fraction of vegetative lipids remained in the biomass residues which was recovered at the back end. The lipid recovery efficiency of the total process for untreated biomass was calculated to be 75.9% which showed approximately 17% improvement for pretreated biomass residues (88.7%). In addition, enzymatic saccharification of pretreated biomass residues recovered > 90% of cellulosic sugars. Based on the average biomass yield of transgenic energycane to be 45 tons/ha (Chiluwal et al., 2018; Knoll et al., 2021; Yang et al., 2018), the calculated potential and actual lipid yields of transgenic energycane were 0.75 tons/ha and 0.66 tons/ha, respectively. The experimental results of the study show that transgenic energycane has the potential to surpass the oil yield of soybean if the metabolic burden due to the expression of transgenes does not have a considerable effect on the biomass yield of the transgenic variety.

References

Chiluwal A, Singh HP, Sainju U, Khanal B, Whitehead WF, Singh BP. 2018. Spacing Effect on Energy Cane Growth, Physiology, and Biomass Yield. *Crop Sci.* **58**:1371–1384. https://onlinelibrary.wiley.com/doi/10.2135/cropsci2017.08.0513.

Clark TJ, Schwender J. 2022. Elucidation of Triacylglycerol Overproduction in the C4 Bioenergy Crop Sorghum bicolor by Constraint-Based Analysis. *Front. Plant Sci.* **13**:1–24. https://www.frontiersin.org/articles/10.3389/fpls.2022.787265/full.

Huang H, Long S, Singh V. 2016. Techno-economic analysis of biodiesel and ethanol co-production from lipid-producing sugarcane. *Biofuels, Bioprod. Biorefining* **10**:299–315. https://onlinelibrary.wiley.com/doi/10.1002/bbb.1640.

Knoll JE, Anderson WF, Missaoui A, Hale A, Hanna WW. 2021. Biomass production and stability of five energycane cultivars at two latitudes in Georgia. *Agrosystems, Geosci. Environ.* **4**:1–14.

Luo G, Cao VD, Kannan B, Liu H, Shanklin J, Altpeter F. 2022. Metabolic engineering of energycane to hyperaccumulate lipids in vegetative biomass. *BMC Biotechnol.* **22**:24. https://doi.org/10.1186/s12896-022-00753-7.

Maitra S, Dien B, Long SP, Singh V. 2021. Development and validation of time-domain 1 H-NMR relaxometry correlation for high-throughput phenotyping method for lipid contents of lignocellulosic feedstocks. *GCB Bioenergy*. https://onlinelibrary.wiley.com/doi/10.1111/gcbb.12841.

Maitra S, Long SP, Singh V. 2022a. Optimizing Chemical-Free Pretreatment for Maximizing Oil/Lipid Recovery From Transgenic Bioenergy Crops and Its Rapid Analysis Using Time Domain-NMR. *Front. Energy Res.* **10**:1–13. https://www.frontiersin.org/articles/10.3389/fenrg.2022.840418/full.

Maitra S, Singh V. 2021. Balancing sugar recovery and inhibitor generation during energycane processing: Coupling cryogenic grinding with hydrothermal pretreatment at low temperatures. *Bioresour. Technol.* **321**:124424. https://linkinghub.elsevier.com/retrieve/pii/S0960852420316989.

Maitra S, Viswanathan MB, Park K, Kannan B, Alfanar SC, McCoy SM, Cahoon EB, Altpeter F, Leakey ADB, Singh V. 2022b. Bioprocessing, Recovery, and Mass Balance of Vegetative Lipids from Metabolically Engineered "Oilcane" Demonstrates Its Potential as an Alternative Feedstock for Drop-In Fuel Production. *ACS Sustain. Chem. Eng.* **10**:16833–16844. https://pubs.acs.org/doi/10.1021/acssuschemeng.2c05327.

Parajuli S, Kannan B, Karan R, Sanahuja G, Liu H, Garcia-Ruiz E, Kumar D, Singh V, Zhao H, Long S, Shanklin J, Altpeter F. 2020. Towards oilcane: Engineering hyperaccumulation of triacylglycerol into sugarcane stems. *GCB Bioenergy* **12**:476–490. http://doi.wiley.com/10.1111/gcbb.12684.

Vanhercke T, Belide S, Taylor MC, El Tahchy A, Okada S, Rolland V, Liu Q, Mitchell M, Shrestha P, Venables I, Ma L, Blundell C, Mathew A, Ziolkowski L, Niesner N, Hussain D, Dong B, Liu G, Godwin ID, Lee J, Rug M, Zhou X-R, Singh SP, Petrie JR. 2019. Up-regulation of lipid biosynthesis increases the oil content in leaves of Sorghum bicolor. *Plant Biotechnol. J.* **17**:220–232.

http://doi.wiley.com/10.1111/pbi.12959.

Yang Y, Wilson LT, Jifon J, Landivar JA, da Silva J, Maeda MM, Wang J, Christensen E. 2018. Energycane growth dynamics and potential early harvest penalties along the Texas Gulf Coast. *Biomass and Bioenergy* **113**:1–14. https://doi.org/10.1016/j.biombioe.2018.03.003.

409e Systems for Single Step Consolidated Bioprocessing of Unprocessed Agri Materials Using Thermophiles

Rajesh Sani, Tanvi Govil, David R. Salem

The United States offers substantial competitive advantages in the drive to replace petroleum-based products and the associated chemical synthesis routes. These advantages include: a very strong agricultural sector with the largest amount of arable land in the world, the largest chemicals and plastics market in the world; an existing industry for bio-based products, and global leadership in bioscience and industrial biotechnology. However, even after decades of investment, releasing sugars from lignocellulosic biomass remains a critical bottleneck, retarding the commercial success of the industries based on the sugar platforms. Against this challenge, non-model microorganisms often hold complex phenotypes that could be valuable for the future of biological processes that underpin innovations for bioplastics, bioenergy, and bioproduct research. This work presents the unique physiology and metabolic capabilities of one such non-model microbe belonging to genus *Geobacillus* sp. to depolymerize unprocessed agri-wastes as the carbon source and produce some key industrial biochemicals. This thermophilic *Geobacillus sp. has been adapted to depolymerize,* an average 63 wt.% of the unprocessed (not subjected to physical, chemical, and enzymatic pretreatment) agri-wastes. A comparative analysis of the transcriptomes of this *Geobacillus* sp. on different carbon sources, including glucose, unprocessed corn stover, cellulose, xylan, and lignin, identified key candidate genes regulating the expression of ligninolytic enzyme genes. Specifically, *Geobacillus* sp. can degrade xylan, amorphous cellulose, starch/pullulan, and assimilate disaccharides such as cellobiose, maltose, melibiose, raffinose, stachyose, sucrose, and mannitol. The monomeric sugars that *Geobacillus* sp. can potentially metabolize are (C6) glucose, galactose, fructose, and (C5) xylose, arabinose, and mannose. For the bioprocessing industries, this study presents new insight into the metabolic routes of carbon capture and transfer in a consolidated bioprocessing *Geobacillus* host, where a single microbe is essentially armored to de-bond cellulose, hemicellulose, and lignin.

409f Green Production of L-Alanyl-L-Glutamine By an Efficient Yeast Biocatalyst Expressing α-Amino Acid Ester Acyltransferase without N-Glycosylation

Yimin Li, Wenjie Yuan

L-Alanyl-L-glutamine (Ala-Gln) is a widely used high value-added dipeptide. An efficient and high-quality biocatalyst plays an essential role in the green production of Ala-Gln. Current yeast biocatalysts expressing α-amino acid ester acyltransferase (SsAet) possessed relatively low activity, which may be attributed to glycosylation of SsAet. In order to promote SsAet activity in yeast, N-linked glycosylation site was first identified as the Asn residue at position 442. Subsequently, we successfully eliminated the negative effect of N-glycosylation on SsAet by the removal of both artificial (α-factor) and the native signal peptide, thereby obtaining a novel yeast biocatalyst K3A1 with significantly improved activity. Besides, the optimal reaction conditions (25°C, pH 8.5, AlaOMe/Gln=1/2) of strain K3A1 were determined to achieve the maximum molar yield and productivity of approximately 80% and 1.74 $g\cdot(L\cdot min)^{-1}$ against the substrate AlaOMe, respectively. Consequently, we established a yeast whole-cell biocatalyst K3A1 to achieve the green production of Ala-Gln in a safe, efficient, and sustainable manner. Our system with a sustainable process provides a prospective approach for commercial Ala-Gln production with property of reusability and product safety in the future.

409g High-Efficient Biobutanol Production from Lignocellulosic Feedstocks By Developing Metabolic and Bioprocessing Engineering Approaches of Clostridium Acetobutylicum

Youduo Wu

Corn stover is evaluated as the most favorable candidate feedstock for lignocellulosic butanol production via microbial acetone-butanol-ethanol (ABE) fermentation by *Clostridium acetobutylicum*. In the presence of pretreatment-derived inhibitors, however, intracellular ATP and NADH involved in cell growth, butanol biosynthesis and stress response, are exceedingly challenged owing to its disrupted sugar transport system. Therefore, there is a necessity to develop effective engineering approach to overcome these limitations for high-efficient lignocellulosic butanol production. *C. acetobutylicum* as a hyper butanol producer could not simultaneously utilize glucose and xylose due to glucose-mediated carbon catabolite repression on xylose metabolism. In this project, metabolic engineering of *C. acetobutylicum* was first performed by heterologous expressions of xylonate biosynthetic enzymes (XylB and XylC) from *Caulobacter crescentus*, which resulted in efficient glucose/xylose co-utilization and cofactor NADH compensation. Efficient strategy for lignocellulosic butanol production was then established integrated with extracellular pH regulation. During batch glucose/xylose culture by the engineered strain, improvements on butanol production and productivity were obtained within 24 h with ~20% and ~200% increases compared to those of the wildtype strain. When using non-detoxified corn stover hydrolysate as substrate, butanol production and productivity were dramatically increased by 3- and 6-fold times, respectively. Pleiotropic roles at global levels were also elaborated to elucidating regulatory mechanisms associated with stress tolerance, biological detoxification as well as central carbon metabolism. Based on these above, the combined metabolic and bioprocess engineering approaches developed in this project not only provide new cues for engineering *C. acetobutylicum*, but contribute to lignocellulose-derived biorefinery and make lignocellulosic butanol production more economically viable and competitive.

409h Antioxidant and Antibacterial Silver Nanoparticles from Agro-Industrial Residues Extracts for Biomedical, Pharmaceutical, Cosmetic and Environmental Applications

Arleth Gualle Brito, Lourdes Orejuela Escobar, Andrea Landazuri, Sebastián Ponce, Patricio Rojas

The field of nanotechnology has revolutionized In the last years, due to the green synthesis of metal nanoparticles with applications in science and engineering. The most versatile nanoparticles are the silver ones, which have optical, electrical, magnetic, chemical, photochemical, catalytic, mechanical, and antimicrobial properties against bacteria, fungi, and viruses. Due to these promising characteristics, they have sought to venture mainly into biomedicine[1]. There are different methods to producing AgNPs, one of them is the chemical synthesis which requires expensive, toxic compounds with negative environmental impact. In contrast, green synthesis is an easier, cheaper and more environmentally friendly methodology because it eliminates the use of chemical substances as reducing/capping agents, replacing them with biological agents [2]. Agro-industrial residues and their extracts can be used for the synthesis of AgNPs, since they possess secondary metabolites as bioactive compounds such as polyphenols, flavonoids, alkaloids, amino acids, that can provide antioxidant, antimicrobial, anti-inflammatory, anti-allergic properties; and can be used for biomedical, pharmaceutical, cosmetic, and environmental applications [3].

Other industries can expand their market to products that can include NPs, such as cosmeceuticals since they need bioactive compounds in their formulations. The AgNPs have a broad spectrum and are well accepted due to their antioxidant, antibacterial, anti-inflammatory and antifungal properties. Nanomaterials are found in many cosmetic products including moisturizer, hair care products, makeup and sunscreen [4]. Nanotechnology also allows for an effective release of ingredients, improving the penetration of substances through the skin barrier. The AgNPs have been tested for the formulation of skin cleansers, lotions, creams, shampoos, toothpastes, and deodorants [5]. Additionally, the anti-inflammatory and anticancer effect of AgNPs allow the development of medicines with specific effects [6]. The antimicrobial activity of AgNPs have major applications in diagnosis, anticancer therapy, diabetes control, biosensor materials, drug delivery systems, among others [7].

According to Singh Prasad et al. (2008) [8], the mechanisms, through which AgNPs exhibited antibacterial properties, were by anchoring to and penetrating the bacterial cell wall and modulating cellular signaling by dephosphorylating putative key peptide substrates on tyrosine residues. AgNps behave in three different ways against Gram-negative bacteria: a) AgNPs (1-10nm) attach to the surface of cell membrane and modify its permeability and respiration; b) AgNPs penetrate inside the bacteria and cause damage to sulfur – and phosphorus – containing molecules such as DNA; c) AgNPs release Ag^+ ions which, additionally, contribute to its bactericidal effect.

The AgNPs' antibacterial properties enables their use as cost-effective disinfectants. AgNPs are used as Ag-based air/water filters [9]. In the environmental sector, one of the sustainable development goals for 2030 is providing clean water. Several costly techniques have been used though years such as filtration, advanced oxidation, reverse osmosis, UV irradiation, etc. to remove different contaminants from water. Water filters should be multifunctional to remove the different water pollutants. Hence, nanomaterials that include AgNPs would have the multipurpose approach required to clean contaminated water and remove chemical compounds and moieties as well as microorganisms. Regarding air disinfection, particles, toxic gases, volatile organic compounds, airborne pathogen bacteria, viruses and fungi, that need to be removed for clean air. Techniques such as germicidal irradiations, photocatalyst oxidation and ozonolysis, which are expensive are applied to clean air. Thus, AgNPs–based nanomaterials from agro-industrial residues can help to control air quality, as effective disinfectant agents.

In this research, the antibacterial activity of agro-industrial residue extracts such as cocoa pod husk (CPH) and avocado seed (AS) and their AgNPs was evaluated. The results revealed that AS-AgNPs exhibited antibacterial properties against pathogenic bacteria such as *E. coli*, while CPH-AgNPs exhibited antibacterial activity against *S. aureus* and *E. coli* [3]. Based on these results, other parameters were evaluated such as total polyphenol content, antioxidant activity, that are reported in this study.

The antioxidant activity was measured using the FRAP method of the extract at a concentration of 5 mg/ml and the AgNPs. Results are expressed in µmol Trolox equivalents/mL. The AS extract contains 661.86 (± 16.78), the CPH extract 418.25 (± 14.46). The AS-AgNP and CPH-AgNP showed an antioxidant activity of 698.147 (± 11.18 714.68 (± 8.85) µmol Trolox equivalents/mL, respectively. Comparing the results obtained from the extracts and their the AgNPs, we can identify that the iron-reducing antioxidant power has an increase in the concentration of bioactive compounds in the AgNPs. A significant 1.7-fold increase in antioxidant activity was observed in CPH-AgNPs than in CPH extract, a similar pattern was also identified in the case of AS-AgNPs but in a smaller proportion 1.05 times more than its extract. The increment in the antioxidant capacity of AgNPs may be due to phytochemicals such as terpenoids, flavonoids and phenolic compounds that the extract contains are adsorbed around the spherical Ag NPs and allow them to interact with the silver ion as inhibitors of singlet oxygen acting as hydrogen donors and reducing agents [10].

Regarding the total polyphenol content of the aqueous extracts, the results show that by the Folin-Ciocalteu method at a concentration of 5mg/ml for AS extract is 0.77 (± 0.088) mg GAE/g extract and CPH extract is 0.474 (± 0.059) mg. GAE/g extract. The same methodology was carried out to determine the polyphenol content for the AgNPs, and the results revealed that the AS-AgNPs contained 0.849 (± 0.017) mg GAE/g extract and CPH-AgNPs showed 0.353 (± 0.010) mg GAE/g extract. These results show that the polyphenols content increase in AS-AgNPs compared to the extracts, these compounds can act as a capping and protective agent around the AgNPs [11]. Thus, CPH extract was identified as having a higher concentration of polyphenols compared to CPH-AgNP. This variation can be attributed to the fact that polyphenols can act as reducing agents from the $AgNO_3$ solution to a colloidal AgNPs solution [12]. The variation in the polyphenol content results may be due the presence of different unique bioactive compounds in extracts that will behave differently as reducing agents when exposed to silver nitrate and LED light.

Based on these results, we can conclude that the use of AgNPs from agro-industrial residues of AS and CPH can be an alternative as antioxidants and antibacterials, since they present particular bioactive characteristics that could be used with specific applications due to their antibacterial and antioxidant capacity that are increased by being included within these nano structures. AgNPs has potential in biomedicine for the coating of biomaterials, medical devices, tissue engineering, diagnostic platforms and therapy against cancer.

This research describes the antioxidant and antibacterial properties as well as polyphenol content of AgNPs from agro-industrial wastes from Ecuador, such as AS and CPH. Additionally, we have evaluated the potential applications that these AgNPs may have in biomedical engineering, cosmetics, pharmaceuticals and environmental sectors. Therefore, AS-AgNPs and CPH-AgNPs are good candidates for new product development, however it requires further studies on cytotoxicity and biodegradability and their impact in the environment.

References

[1] P. Mejía and J. Ulloa, "'Estudio de la Inclusión Educativa para los niños y niñas de edad Preescolar,'" UNIVERSIDAD DE CUENCA, 2010.

[2] M. A. Huq, M. Ashrafudoulla, M. M. Rahman, S. R. Balusamy, and S. Akter, "Green Synthesis and Potential Antibacterial Applications of Bioactive Silver Nanoparticles: A Review," *Polymers (Basel).*, vol. 14, no. 4, pp. 1–22, 2022, doi: 10.3390/polym14040742.

[3] A. Cañadas *et al.*, "Green Synthesis of Antibacterial Silver Nanocolloids with Agroindustrial Waste Extracts, Assisted by LED Light," *Colloids and Interfaces*, vol. 6, no. 4, p. 74, Dec. 2022, doi: 10.3390/colloids6040074.

[4] S. Gajbhiye and S. Sakharwade, "Silver Nanoparticles in Cosmetics," *J. Cosmet. Dermatological Sci. Appl.*, vol. 06, no. 01, pp. 48–53, 2016, doi: 10.4236/jcdsa.2016.61007.

[5] W. T. J. Ong and K. L. Nyam, "Evaluation of silver nanoparticles in cosmeceutical and potential biosafety complications," *Saudi J. Biol. Sci.*, vol. 29, no. 4, pp. 2085–2094, 2022, doi: 10.1016/j.sjbs.2022.01.035.

[6] S. Agrawal, M. Bhatt, S. K. Rai, A. Bhatt, P. Dangwal, and P. K. Agrawal, "Silver nanoparticles and its potential applications: A review," *J. Pharmacogn. Phytochem.*, vol. 7, no. 2, pp. 930–937, 2018.

[7] P. Mathur, S. Jha, S. Ramteke, and N. K. Jain, "Pharmaceutical aspects of silver nanoparticles," *Artif. Cells, Nanomedicine, Biotechnol.*, vol. 46, no. sup1, pp. 115–126, Oct. 2018, doi: 10.1080/21691401.2017.1414825.

[8] S. Prasad, "Nanotechnology in Medicine and Antibacterial Effect of," vol. 3, no. 3, pp. 115–122, 2008.

[9] S. P. Deshmukh, S. M. Patil, S. B. Mullani, and S. D. Delekar, "Silver nanoparticles as an effective disinfectant: A review," *Mater. Sci. Eng. C*, vol. 97, pp. 954–965, Apr. 2019, doi: 10.1016/j.msec.2018.12.102.

[10] Z. Bedlovicová, I. Strapác, M. Baláž, and A. Salayová, "A Brief Overview on Antioxidant Activity," *Molecules*, pp. 1–24, 2020.

[11] S. Salari, S. E. Bahabadi, A. Samzadeh-Kermani, and F. Yosefzaei, "In-vitro evaluation of antioxidant and antibacterial potential of green synthesized silver nanoparticles using prosopis farcta fruit extract," *Iran. J. Pharm. Res.*, vol. 18, no. 1, pp. 430–445, 2019.

[12] P. K. Tyagi *et al.*, "Ascorbic Acid and Polyphenols Mediated Green Synthesis of Silver Nanoparticles from Tagetes erecta L. Aqueous Leaf Extract and Studied Their Antioxidant Properties," *J. Nanomater.*, vol. 2021, 2021, doi: 10.1155/2021/6515419.

Methodology

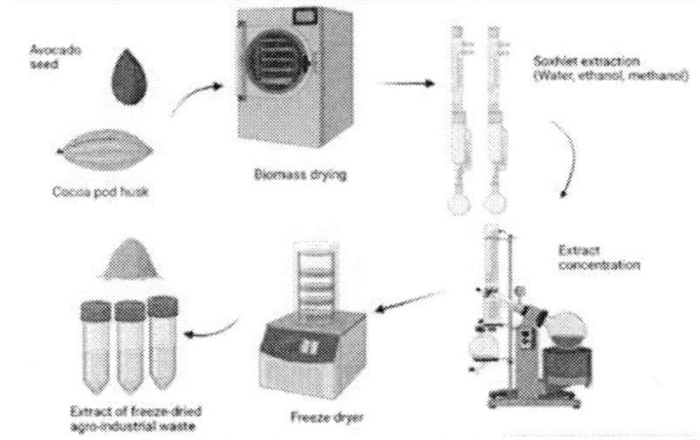

Fig 1. Process for obtaining extracts of agro-industrial residues

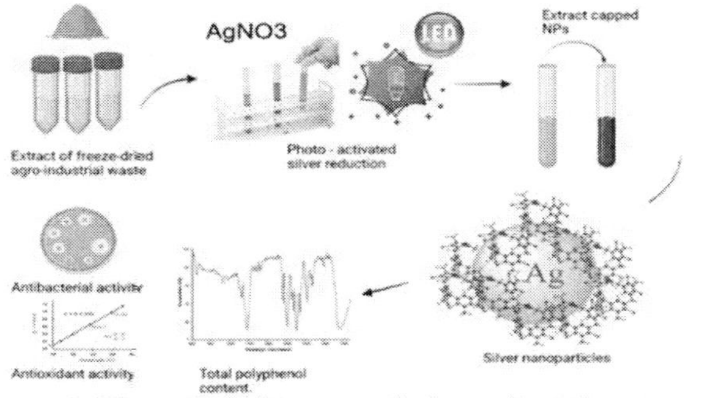

Fig 2. Green synthesis of silver nanoparticles from agro-industrial extracts

Results

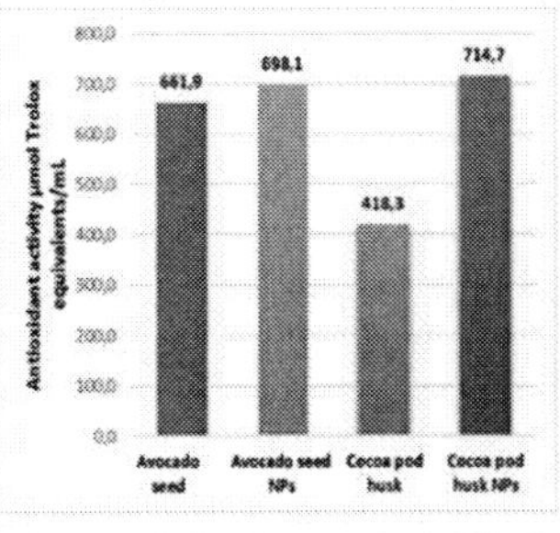

Fig 3. Antioxidant activity of the extract and AgNPs

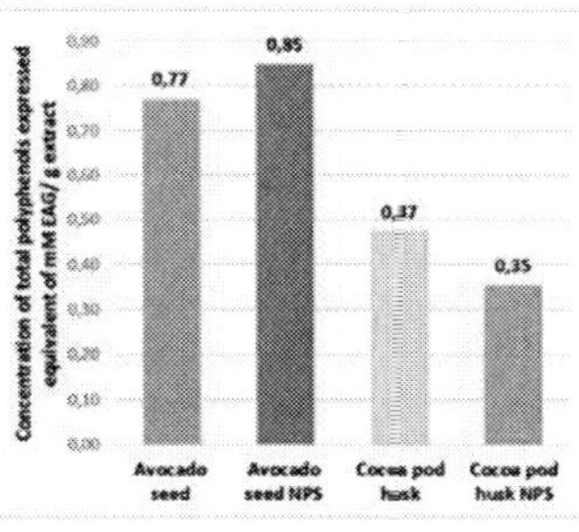

Fig 4. Total polyphenol content of the extract and AgNPs

Table 1. Antibacterial activity of aqueous extracts and silver nanoparticles from agro-industrial residue extracts against bacterial microorganisms.

	Sample	Staphylococcus aureus Inhibition zone mm (SD)	Escherichia coli Inhibition zone mm (SD)
1	AS extract	0.0	0.0
2	AS-NPs	0.0	10.51 ± 0.087
3	CPH extract	0.0	0.0
4	CPH-NPs	11.51 ± 0.005	13.8 ± 0.025
Negative control	Distilled water	0.0	0.0
Positive control	Gentamicin Discs 10 µg	21.18 ± 0.055	17.48 ± 0.029

Applications

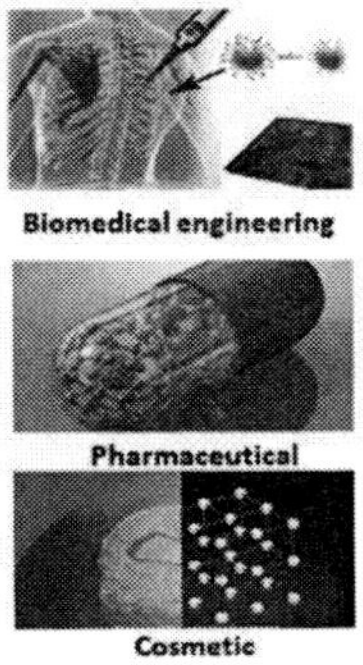

446a Nature-Based Solutions: Food Waste-Derived Carbon Materials and Its Application in Air-Breathing Energy Storage

Mengyao Gao

The sluggish kinetics of air electrodes prohibits the practical applications of metal-air batteries. At present, preparing single-atom catalysts confined in S, N-doped porous carbon (NPC) with a high surface area remains a challenge. Inspire by nature, waste from food and agriculture has significantly attracted our attention, e. g. fish scale, and shrimp shell where the inorganic part (hydroxyphalite, $CaCO_3$) could act as template, while organic part (collagen fiber, chitin fiber) is responsible for the carbon skeleton. The as-prepared NPC with a hierarchical structure would not only perform as oxygen reduction reaction (ORR) catalyst, but provide abundant active sites for the charge-discharge process of air-breathing batteries. This study creates a pathway for high-performance ORR single atomic electrocatalysts for air-breathing batteries applications and provides opportunities to convert biowaste materials into commercial opportunities. We will further deploy the waste-derived catalyst to Na-CO_2 batteries to contribute the Net Zero Emissions. Our research team aims to pave a new avenue for the implementation of applying nature-based solutions (NbS) and green circular technologies towards responsible consumption and production in the scope of the Sustainable Development Goals (SDGs).

446b Carbon-Negative Biomaterials Via Engineered Composting with Carbon Capture

Joe Sagues, Ethan Woods, Vanessa Rondon Berrio, Nicolas Clauser

Biodegradable products are advancing rapidly due to the negative consequences associated with disposal of traditional, non-biodegradable products. Composting is advancing rapidly in parallel to biodegradable products due to the need for efficient and economical disposal of the new biodegradable products. Although composting offers many benefits, the net greenhouse gas emissions associated with compostable products are often slightly better, and sometimes worse, than traditional products. The opportunity exists to capture and sequester CO2 emitted from composting, thereby making compostable products truly carbon negative. For the first time, we have engineered a method of composting which produces a concentrated stream of carbon dioxide for subsequent capture and sequestration with a relatively low capital and infrastructure intensity. In collaboration with NCSU's compost facility, we have rapidly composted biological materials with semi-continuous collection of gas containing 20 – 95% CO2 using a novel closed reactor design. Such a technology has the potential to remove > 1 billion tons of CO2 per year from the atmosphere if bioproducts, including bioplastics and biofiber, significantly penetrate existing markets.

446c Biowaste-Derived, Nanocomposite-Based Slow-Release Fertilizer Encapsulated with a Super Thin Coating Layer

Jiaxuan Liu, Zhaohui Tong

Nutrient management is an important issue in fertilizer study to enhance crop yield with minimal environmental impact. Achieving the right balance between the amount of fertilizer applied for the nutrient absorption and minimizing harmful effects like soil degradation due to nitrite accumulation is crucial. Slow-release fertilizers (SRFs) have received significant attention as they offer a promising solution for crop nutrient management by improving nutrient efficiency, fostering crop growth, and reducing fertilizer waste and pollution through optimization and control of the extended release period.

However, current SRFs suffer from challenges in high release speed and high cost. To tackle these challenges, this study proposes a novel fabrication method to prepare bio-based SRFs with a controllable fertilizer release function and super thin coating layer to reduce production costs. A nanoclay-based bio-polyurethane membrane is synthesized from modified bentonite clay and bio-polyol derived from biomass residue. The modification of bentonite clay results in a fully exfoliated nano clay distributed in the polymer matrix of the membrane, followed by the addition of polyol that enables in-situ reaction and formation of a denser cross-linked structure. Ultimately, the synthesized nanoclay-based slow-release fertilizers (CSRFs) demonstrated excellent control over nitrogen release for a time period of more than one month and promoted plant growth of soybean in a hydroponic system. The soybean showed improved plant height, leaf greenness, shoot biomass, and yield compared to traditional fertilizers.

Additionally, this work provides a detailed examination of the synthesis of modified nanoclay, the encapsulation of urea fertilizer by bio-nanocomposite, and the study of the dynamic nutrient release profile. The sustainable CSRFs fertilizers from bio-waste provide a cost-effective solution for optimizing the nutrient release profile, demonstrating the potential to substitute conventional fertilizers.

446e Synergistic Enhancement of Barrier, Aging Resistance, and Hydrolysis for Biodegradable Copolyesters Based on Poly(butylene adipate terephthalate) and Polyglycolic Acid

Bingxue Jiang, Yisong Wang, Wen-Jun Wang, Pingwei Liu, Xuan Yang, Bo-Geng Li

As one of the most promising biodegradable plastics, PBAT has attracted immense attention in the fields of packaging and mulch films. However, limitations of degradation performance in aqueous environment and mechanical properties hinder its further development. In this work, we focused on chain structure design, introduced glycolic acid units into the PBAT backbone, and synthesized a series of PBAT-*co*-PGA copolyesters with different compositions and sequence structures by melt copolycondensation. The number-average sequence length of the copolymer units can be varied while keeping its molar content unchanged by adjusting the molar ratio of copolymer monomer in PBAT-*co*-PGA. PBAT-*co*-PGA exhibited higher Young's modulus (88 MPa) and tensile strength (35 MPa) due to the better crystallization ability compared to PBAT with the same content of terephthalic acid. We found that incorporation of more terephthalic acid and glycolic acid resulted in higher glass-transition temperature and a rigid chain structure, which is beneficial to improve the water vapor barrier properties of copolyesters. Additionally, PBAT-*co*-PGA copolyesters showed enhanced hydrolytic degradablity in lipase and alkaline solutions without sacrificing the aging resistance. The synthetic strategy proposed here endow the PBAT-*co*-PGA copolyesters with a wider chain structure regulation window than PBAT, which makes it have excellent mechanical, H_2O barrier, and ageing-resistant properties, also accelerates the enzymatic and non-enzymatic hydrolysis of copolyester.

446g Production of Polyhydroxybutyrate from Forest Residue Using a Chemical-Free Pretreatment

Md Shahadat Hossain, Bandaru V. Ramarao, Obste Therasme, Timothy A. Volk, Vinod Kumar, Deepak Kumar

Polyhydroxybutyrate (PHB) is a class of biodegradable and biocompatible biopolymers, which is one of the alternatives to degradation-recalcitrant petroleum-based plastics. PHB production and application have been limited on a commercial scale despite considerable interest and effort, mostly because of high production costs. Abundant and low-cost lignocellulosic biomass can be used as carbon sources to lower the overall bioplastic production cost. Forest residue biomass (FRB) generated from harvesting operations, including tops, branches and low-grade wood material, is one of the lowest cost and abundantly available feedstock to produce sugars that can be subsequently fermented to produce PHB. However, FRB has a recalcitrance structure which makes a pretreatment step necessary to improve the biomass hydrolysis efficiency. Conventional pretreatment processes use acids or alkalis at high temperatures and pressures, which results in high energy use and the formation of intermediate compounds inhibitory to fermentative microorganisms. Alternatively, biomass can be pretreated using a sequential hydrothermal pretreatment-disk refining process and obtain hydrolysis yields similar to those from chemically pretreated biomass. The objective of this work is to optimize the two-stage hydrothermal pretreatment conditions to maximize sugar yields for PHB production. The pretreatment experiments were performed using a central composite design for three variables: temperature (160-200 °C), time (10-20 min), and solid loading (10-20 %). Hydrothermally pretreated biomass was disk milled and hydrolyzed using commercial cellulase and hemicellulase enzymes. The sugar yields of pretreated biomass were higher than untreated biomass for all pretreatment conditions. The high severity pretreatment (213.6°C, 15 min, and 15% solid loading) resulted in more than 95% glucan conversion. However, pretreatment at these conditions resulted in only 21% hemicellulose recovery after hydrolysis, due to sugar degradation. The pretreatment conditions of 194.8°C, 12.9 min, and 13.4% solid loading were found optimum for maximizing the total sugar conversions (87.8%). The hydrolysate obtained from the hydrolysis of biomass pretreated at these optimized conditions will be evaluated as the carbon source for PHB synthesis by Recombinant *Escherichia coli* LSBJ and the results will be compared with the PHB yield from pure sugars.

446h Development and Characterization of Biobased Biodegradable Sprayable Mulching Materials to Control the Weeds in Agricultural Crops

Muhammad Ehtasham Akram, Mark Wilkins, Loren Isom, Ozan Ciftci

Polyethylene plastic mulch (PPM) films have been employed in high-value and specialty crop production as a physical barrier to weed growth since the 1950s. Over the last decade, due to the environmental problems associated with PPM, biodegradable plastic mulch (BPM) films have started to replace PPM films. However, concerns have arisen regarding the control of BPM film deposition, removal, and decomposition. Biobased sprayable mulch (BSM) films are a promising substitute for BPM films. A series of BSM films was developed using locally available biomaterials, including corn starch (SC), potato starch (PS), chicken feather keratin hydrolysate (CFKH), corn zein (CZ), isolated soy protein (ISP), pea protein (PP), and lignin. Optimum BSM film-forming conditions of protein (CZ, ISP, and chicken feather keratin), starch, and plasticizer agent (glycerol) mixtures were determined by evaluating different protein/starch and protein/glycerol ratios. The isolation of keratin from chicken feathers was carried out by dissolving it in sodium hydroxide (NaOH) solution. The chemical and physical properties of BSM film-forming solutions were analyzed (pH, total dissolved solids (TDS), conductivity, salinity, viscosity) and transferred to petri dishes and dried at 40°C for 24 hours. The dried BSM films were peeled off from the petri dishes for mechanical and other (FTIR, TGA, SEM) characterization, water absorption, and percentage of mass loss in water. The maximum tensile strength observed was 0.013 MPa, with an elongation at break of 25.7% and a young's modulus of 0.269 MPa. Finally, two of the BSM film-forming solutions that showed higher mechanical strength and elongation at break were tested under greenhouse conditions.

516a Hydrogen Production Via Biomass Gasification Using Calcium Ferrite Based Chemical Looping

Shekhar Shinde, Rushikesh Joshi, Liang-Shih Fan

The current energy scenario demands extensive research to efficiently produce clean fuels such as hydrogen from renewable energy sources. Hydrogen has proved to be a clean fuel as it produces no greenhouse emissions during its combustion. Traditional hydrogen production has relied on the gasification of fossil fuels through processes such as Coal gasification, Steam Methane Reforming, Autothermal Reforming, etc. However, with the increasing demand for hydrogen and depleting fossil fuels, it is essential to shift hydrogen production towards greener processes that use renewable energy sources. Several attempts have been made to achieve this goal, but these processes have reported lower efficiencies and higher production costs. However, a chemical looping process involving biomass gasification is an excellent alternative for hydrogen production, where the biomass is converted to syngas that can be integrated with the conventionally established pathways to produce molecular hydrogen.

This study aims to produce hydrogen through a chemical looping process that utilizes Calcium Ferrite as an oxygen carrier (OC). The proposed process has a moving bed reactor wherein the biomass and the OC particles are fed co-currently to selectively produce partial oxidation products.

Calcium Ferrite is a redox material with potentially no adverse effects on the environment, making it an attractive alternative for use in chemical looping technology. Due to its fast ionic oxygen transport mechanism, Calcium Ferrite has shown high activity owing to its brownmillerite structure. The thermodynamic properties of calcium ferrite aid in the controlled oxidation of biomass. This favors the production of partial oxidation products of biomass, i.e., Syngas, while the calcium ferrite material itself undergoes reduction to lower oxidation states. Another peculiar advantage of calcium ferrite is its ability to completely regenerate in CO_2 or steam, eliminating the air oxidation step required in conventional chemical looping systems and thus reducing the associated costs. The regeneration step produces carbon monoxide or hydrogen based on the regeneration media, which can be further utilized to tailor the syngas produced.

In this system, the biomass and OC particles are fed from the top of the reactor, wherein they flow co-currently downwards. The top portion of the moving bed reactor contains oxidized oxygen carrier particles that react with the freshly added biomass. These particles get reduced on progressing further down through the reactor as they continuously react with biomass volatiles. An enhancer gas is injected from the top to gasify the char. The product syngas is obtained at the bottom of the reactor and sent to the water-gas shift reactor, followed by pressure swing adsorption and acid gas removal unit for hydrogen production. The reduced oxygen carriers then enter the regeneration step, wherein they are oxidized by steam to produce additional hydrogen, along with the complete regeneration of the oxygen carriers.

Initially, process simulations are performed to optimize the system performance, and consequently, the reaction conditions are determined. The experiments are carried out in three significant sets- simulated moving bed experiments to study the feasibility, moving bed bench-scale experiments, and TGA experiments to study regeneration characteristics. The simulated moving bed consists of two layers of OC particles: the lower layer of reduced particles and the upper layer of oxidized OC particles. This arrangement is made to simulate a moving bed system closely. The biomass is fed using a lock hopper system from the top of the reactor. In a moving bed system, the biomass is mixed with the OC particles and fed from the top of the reactor at a controlled rate. The TGA experiments are designed to study the regeneration characteristics of reduced OC particles in CO_2 and steam environments. The regeneration step is also replicated in the moving bed system wherein the reduced OC particles and steam move counter-currently for complete regeneration of OC particles. High-purity syngas is obtained with a high H_2/CO ratio and fed to the water-gas shift reactor to produce high-purity hydrogen. Hydrogen generation from both steps- Biomass gasification and OC regeneration, can be accounted for a high H_2/CO ratio.

Various biomass feedstocks have been studied in the system, and high-purity syngas is obtained in each case, proving the system's applicability to different feedstocks. Further selectivity and conversion are improved by adding dopants to the OC material.

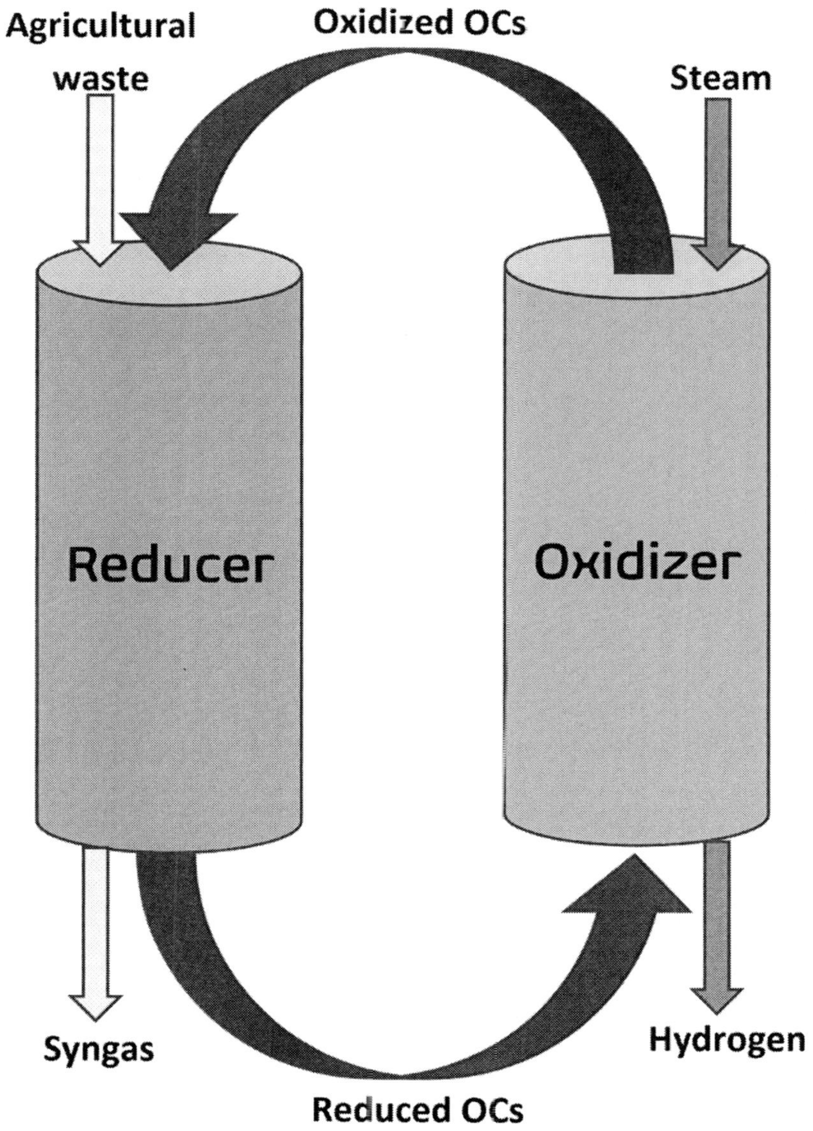

516b Alkaline Thermal Treatment of Challenging Waste Biomass Feedstocks for *in-Situ* H_2 Production with Tandem Carbon Sequestration

Jonah Williams, Ah-Hyung Park

Bioenergy with carbon capture and storage (BECCS) is a method of carbon dioxide removal which will be a critical component of CO_2 removal and storage from the atmosphere. In the IPCC-AR5 report, it was concluded that warming below 2°C could not be achieved if BECCS deployment is limited (*high confidence*). This is because BECCS is theoretically carbon-negative as CO_2 is captured by both biomass growth and the coupled downstream CO_2 capture. The thermochemical conversion of biomass (e.g., gasification) are among the most prominent methods for BECCS as they combine proven technologies (e.g., thermal gasifiers) with emerging carbon capture systems. Unfortunately, they are often hampered by high temperatures and pressures, feedstock purity requirements, pure oxidant demand, and low actualized CO_2 capture ability. Alkaline thermal treatment (ATT) is a promising and robust BECCS method which can run at mild temperatures and pressures (<500°C, 1 bar) and produces carbon-free high-purity H_2 (>90% v/v) in a single reactor. Hydroxide salts are utilized during the ATT reaction (e.g., NaOH, KOH, LiOH, etc.), facilitating the degradation and conversion of organic matter. Tandemly, carbon from the biomass is captured in the form of stable carbonate salts (e.g., Na_2CO_3). In this study, we report on the H_2 production and CO_2 capture potential of a wide variety of difficult-to-process biomass sources using alkaline thermal treatment, including wet and salty seaweeds, construction and demolition organics, and plastics commingled with marine wastes. Characterization of the reaction intermediates and products is performed to better elucidate the underlying reactive mechanisms. Gas-phase reforming catalysts, such as Ni/ZrO_2, are utilized to boost H_2-yields and increase the degree of energy recovery from the biomass. Finally, regeneration methods to recover the spent alkaline material for further conversion are considered, including metathesis with Ca-rich industrial wastes or molten salt electrolysis for carbonate electrosplitting.

516c Surfactant Assisted Catalytic Hydrothermal Liquefaction (CHTL) of Corn Stover for Fuels and Value-Added Products

Bharathkiran Maddipudi, Rajesh Shende, Khang Huynh, Anuradha Shende

Worldwide, as corn crop yield is more than 1 bty, significant amount of corn stover is available for bioprocessing. In this study, we investigated hydrothermal liquefaction (HTL) of corn stover at 225°C to 300 °C and 800- 1300 psi (system autogenous pressure) using different catalytic materials and surfactants. Catalytic hydrothermal liquefaction (CHTL) of corn stover and derived unhydrolyzed solids (UHS) was performed at different processing conditions of temperature, initial nitrogen pressure, biomass to water ratio, and catalyst loading. Presence of a surfactant is known to change the wettability, reduce the hydrophilicity and increase interfacial surface area, which can influence characteristics of hydrochar as well as distribution of liquified products, mainly the C1-C3 carboxylic acids. HTL processing resulted in product gas, aqueous biocrude, heavy bio-oil (HBO) and hydrochar. The aqueous biocrude was characterized by the TOC analyzer and gas-chromatography mass spectrometry (GC-MS) to understand extent of liquefied carbon and oxygenated hydrocarbons, respectively. Both hydrochar and HBO were analyzed by the elemental analyzer and HHVs were estimated based on C, H, N and O content. Liquid - liquid extraction methodology was developed to separate phenol/substituted phenols from aqueous biocrude. HTL processing, separation methodologies, process engineering aspects and scale-up strategies will be presented for the conversion of corn stover and UHS into value-added products.

516d Sustainable Aviation Fuel Production through Co-Hydroprocessing of Catalytic Fast Pyrolysis Oil with Petroleum Streams

Xiaolin Chen, Kristiina Iisa, Kellene A. Orton, Calvin Mukarakate, Michael B. Griffin

Catalytic fast pyrolysis (CFP) is an efficient way to produce stabilized low-oxygen biogenic oil from lignocellulosic biomass, which is an abundant and domestically available feedstock and does not contribute to the competition between food and energy. Hydrotreating can further convert CFP oil into various renewable transportation fuels. Due to the growth in electrification of the light-duty transportation sectors a large distillate hydrotreating capacity is available and ready for use. The studies on co-hydrotreating of CFP oil with petroleum streams can provide routes to incorporate biogenic oil into fuel pool with minimal capital costs. Previously, renewable diesel with a high indicated cetane number of 45 was produced through co-hydrotreating of CFP oil with straight-run diesel (SRD) under a typical refinery SRD hydrotreating condition, which transferred up to 97% of biogenic carbon from CFP oil into final renewable diesel product. In this project, co-hydrotreating of CFP oil with SRD and co-hydrocracking of CFP oil with a mixture of vacuum gas oil (VGO) and light cycle oil (LCO) were conducted. CFP oil with 22% oxygen on dry basis was produced from pine in a bench-scale fluidized bed reactor over an HZSM-5 catalyst, after which the CFP oil was hydroprocessed in a continuous trickle-bed reactor over commercial catalysts. The chemical composition of the co-hydroprocessed products was characterized by elemental analysis, simulated distillation, GCxGC-TOFMS-FID and GC-VUV. The co-hydroprocessed products were fractionated by distillation into fuel cuts, and selected fuel properties (density, viscosity, heating value, freeze point, indicated cetane number) for the co-hydroprocessed sustainable aviation fuel (SAF) product were evaluated and compared with ASTM requirements. Stand-alone hydroprocessing of the petroleum products was studied under the same condition as comparisons. Producing SAF from co-hydroprocessing of CFP oil with petroleum streams has great potential in mitigating GHG emissions at the same time reducing the fuel production costs by making the most of the existing refinery infrastructure, workforce, and institutional knowledge.

516e Automated Reaction Network Generation and Kinetic Modeling for Fast Pyrolysis of Lignin with Model Compounds

Tanzina Azad, Maria Auad, Thomas Elder, Andrew Adamczyk

Lignin is one of three major structural components of the lignocellulosic biomass and the second most abundant natural polymer. Due to its highly heterogenous molecular structure, lignin contributes most of the recalcitrance to biomass valorization. Fast pyrolysis is a promising thermochemical process for lignin valorization, allowing less char formation. However, the knowledge of the reaction mechanism for this process still needs to be completed, and many information gaps exist among the experimental and computational studies on this topic. To address it, we employ the concept of automated reaction network generation and kinetic modeling to study lignin pyrolysis using the model compounds. The computerized reaction network generation approach is a beneficial and realistic choice to do microkinetic modeling for such a complex and large reactive system. In this work, we constructed detailed reaction mechanisms for the fast pyrolysis process of anisole and guaiacol. These oxygenated aromatic compounds are the pyrolysate from lignin pyrolysis and serve as model lignin compounds. These compounds are also used as surrogates for lignin-derived primary tar. Therefore, a better understanding of gas-phase reactions among the species that evolved during the pyrolysis of these model compounds also helps better understand other thermal conversion processes, including gasification and combustion. These model compounds have been experimentally investigated, and specific product spectra have been reported in the literature. However, many missing links exist for the pyrolysis process for both model compounds. We employed the automated reaction network generation approach using these two model compounds and performed extensive reaction pathway analysis that would be insightful for bridging the existing information gaps. The resulting simulation from our work shows good agreement with the experimentally reported concentration profiles of pyrolysis products. We checked the effect of pyrolysis temperature on the generated reaction networks and analyzed the pathways for forming major products. We identified several areas of improvement so that this relatively less conventional approach to microkinetic modeling can be further used for models with similar complexity. Therefore, this work is the foundation for the successive model extension to automated reaction network generation and microkinetic modeling for lignin pyrolysis.

516f Investigating C-C Coupling Pathways for the Sustainable Production of Liquid Fuels

Laura Paz Herrera, Randy Cortright, J. Will Medlin

The search for renewable alternatives to current industrial products, processes, and technologies has been a priority for the field of catalysis during the past decades. In particular, research efforts have focused on utilizing abundant, renewable resources available in the United States, such as biomass, and transforming them into fuels and value-added products in a sustainable manner. Crude oil, a non-renewable unrefined petroleum feedstock subject to depletion, has served as the principal energy source for processing and refinement into transportation fuels. Thus, both environmental and economic reasons have driven the development of biological, catalytic, and thermochemical pathways for deconstruction, conversion, and further upgrading of biomass into advanced materials The thermochemical pathway for conversion typically involves hydrogenation, dehydration, hydrodeoxygenation, isomerization, and carbon-carbon coupling reactions. Carbon-carbon coupling reactions are key in the catalytic conversion step of the process, as they enable the construction of complex organic molecules by forming new carbon-carbon bonds and can be rate limiting under a variety of conditions. In this work, propionaldehyde and ethylene were studied as model compounds representative of cellulosic-derived biomass deconstruction products. Conversion of these compounds into longer carbon-chain structures over aluminosilicate materials via three coupling routes, including aldol condensation, Prins reaction, and oligomerization was studied. The effects of ethylene : propionaldehyde feed ratio, reaction temperature, and catalyst type (H-Al-MCM-41, H-ZSM-5, SiO_2-Al_2O_3) were investigated. Our findings suggest that propionaldehyde self-aldol condensation to form 2-methyl-2-pentenal is the predominant pathway, followed by the acid-catalyzed Prins reaction between ethylene and propionaldehyde to form 1,3-pentadiene.

516g Pyrolysis Oil: A Promising Anode Precursor for Lithium-Ion Batteries

Shaikat Chandra Dey, Lilian Lower, Joe Sagues, Bertrand Tremolet de Villers, Sang-Don Han, Mark R. Nimlos, Steve Kelley, Sunkyu Park

Graphite demand is increasing rapidly due to the popularity of electric vehicles (EVs). Lithium-ion batteries (LIBs) are the power source of EVs and the anode of LIBs is made of graphite. This study aims to develop and optimize a novel process for the production of battery-grade graphite from a renewable precursor, pyrolysis oil, using iron as the graphitization catalyst. Catalytic graphitization was carried out under relatively moderate conditions (1500 °C, 20 hours) to reduce costs and environmental impacts as compared to incumbent graphitization technologies. After graphitization, the residual catalyst was separated by washing the graphite-iron mixture with hydrochloric acid under reflux. The efficiency of iron removal was studied using both X-ray fluorescenc (XRF) and inductively coupled plasma (ICP) techniques. To evaluate the quality of graphite, X-ray diffraction (XRD), Raman spectroscopy, and high-resolution transmission electron microscopy (HR-TEM) were applied as the preliminary screeing tools. The optimized graphite sample demonstrated excellent reversible capacity (>350 mAh/g) with minimal capacity loss over 100 cycles and a high initial Coulombic efficiency (90%) when applied as a lithium-ion battery anode, thereby making this technology potentially competitive with commercially available battery-grade graphite.

516h Optimization and Evaluation of the Distribution of Fischer-Tropsch Products over a Cobalt-Based Catalyst Utilizing Design Expert Software

Roick Chikati, Joshua Gorimbo, Diakanua Nkazi

Modelling biomass to liquid via the Fischer-Tropsch synthesis (FTS) system allows researchers to investigate the most efficient parameters while running the system under optimal conditions. As part of the design of experiments (DOE) procedure, a special data simulation method based on response surface methodology (RSM) is utilized to thoroughly analyze the impact of operating circumstances. The objective of this study was to examine the factors that affect the production of C_1, C_2-C_4, and C_{5+} in FTS process, and then optimize the critical factors utilising factorial design and response surface techniques. The parameters evaluated were reaction temperature, reaction pressure and the crystallite size of cobalt. The effects of these factors and their potential for synergy were explored simultaneously using multivariate DOE, with the yield of different hydrocarbon composition selectivity's as the measured responses. In the concept generation phase, optimization was based on the literature consulted, which proved to be an effective method for determining the optimization parameters. The detailed conceptual design included the generation of models using statistical methods and response surface models. Finally, the optimized design was validated using catalysts and parameters obtained during the optimization process, and this were compared to the output recorded in the theoretical modelling. The optimized parameters resulted in performance consistency, with the theoretical model for each group of hydrocarbons being validated by actual experiments. The established models were seen to characterize hydrocarbon distributions accurately and repeatedly over a wide range of reaction conditions (200-270 °C, 5-20 Bar, and 3-26 nm) using a cobalt-based catalyst. According to the detailed quantitative models developed, for higher C_{5+} production, 220 °C, 10 barg and 11nm (cobalt crystallite) benchmark parameters were set to produce 19.3 % C_1, 11.4% C_2-C_4 and 69% C_{5+} selectivity's. Comparative analysis showed a 1.9%, 3.9% and 0.3% percentage difference between the theoretical output and the actual output of C_1, C_2-C_4 and C_{5+}, respectively.

578a Producing a Portfolio of Commodity Chemicals from Lignin Bound *P*-Hydroxybenzoate

Steven Karlen, Vitaliy Tymokhin, Canan Sener, Justin Mobley, John Ralph

Finding abundant renewable sources for common commodity chemicals and pharmaceuticals is one of the great challenges of the day. One potential chemical is *p*-hydroxybenzoic acid which is found in Salicaceae trees (*e.g.*, poplar, willow, aspen, and cottonwood) and in Arecaceae (*e.g.*, oil palm and date palms) and can be up to 3wt% of the woody biomass. Though a commodity chemical on its own, *p*-hydroxybenzoic acid can be converted to a portfolio of high value commodity chemicals that are precursors to textiles, pigments, resins, and biodegradable plastics. We constrained our process to fit the following criteria, 1) Start from woody biomass or isolated lignin. 3) Produce 4-aminophenol and paracetamol in high yield. 3) Utilize inexpensive isolation and purification techniques. and 4) Produce a pure colorless solid product. Here in we will report our progress on developing a three-step process for the making paracetamol from lignin bound *p*-hydroxybenzoates via the Hoffman rearrangement. We cover some of the issues associated with the process and our strategies to address those challenges.

578b Advances in the Biomass-to-Bioproduct Processing Chain to Produce 2-Pyrone-4,6-Dicarboxylic Acid Via Reductive Catalytic Fractionation Followed By Microbial Funneling

Canan Sener, Miguel Perez, German Umana, Steven Karlen, Christos Maravelias, John Ralph, Timothy J. Donohue, Daniel R. Noguera

Catalytic, enzymatic, and microbial conversion are used in different stages of the new generation biorefineries. In an integrated biomass-to-bioproduct processing chain, these processes are interrelated, and this makes the optimization of the overall processing chain challenging. The biggest pitfall in developing these processes is the mismatch between the goals of catalytic fractionation of biomass and subsequent biological/enzymatic upgrading.

In our previous work, we used chemical and biological upgrading in tandem to extract greater value from the lignin fraction by converting a fraction of lignin to high value 2-pyrone-4,6-dicarboxylic acid (PDC). Reductive Catalytic Fractionation (RCF) was used to liberate lignin from the biomass and simultaneously depolymerize the lignin through catalytic hydrogenolysis. Using an engineered strain of *Novosphingobium aromaticivorans* DSM12444, the RCF product mixture can be upgraded to PDC. By applying RCF to the poplar with Pd/C as catalyst and methanol as solvent; we were able to have increased monomer/oligomer product yield and the PDC yield on a per kg of biomass basis compared to a stepwise lignin isolation followed by hydrogenolysis of fractionated lignin. Furthermore, the sugar stream of the lignocellulosic biomass which was preserved after RCF process, was subjected to enzymatic and microbial digestion to produce liquid fuels. Combined, we produced value added products from both the lignin (phenolics) and polysaccharide fractions.

Technoeconomic analyses of our processing chain indicated that the RCF stage is costly because of the high-pressure reaction conditions due to low boiling point of methanol. Changing the solvent from methanol to a higher boiling point one will not only improve the process economics but also improve the safety of process operation. For this purpose, ethylene glycol, ethanol and ethanol-water mixture were tested. In our processing chain, using these solvents with reduced reaction pressure have shown that similar lignin monomer yields with methanol as solvent can be achieved. Initial experiments showed that these solvents are not detrimental to microbial upgrading and similar PDC yields could be obtained. Further process optimization was performed by varying the reaction time, solid concentration, biomass/catalyst ratio, and the effect of these changes on the microbial funneling to PDC was determined.

578c Towards the Valorization of Waste-Stream Feedstocks into Value-Added Chemicals Using Genetically Engineered Yeasts

Mohamed Nasr, Radhakrishnan Mahadevan

The fields of synthetic biology and metabolic engineering aim at reprogramming microbes to be used in a multitude of applications. To this end, we are engineering yeasts towards the valorization of potato starch obtained from food waste streams to produce chemicals of industrial interest. Ongoing work includes engineering the model yeast *Saccharomyces cerevisiae* for digestion and utilization of pure potato starch. This will be followed by strain and enzyme engineering efforts for improving efficiency of starch digestion such as enhancing enzyme secretion, in addition to enzyme copy number variation and directed evolution. Furthermore, the engineered strains will be characterized with a minimally processed potato-based feedstock obtained from waste streams as the main carbon source for sustaining cell growth. Finally, in an effort to contribute towards a meaningful circular bioeconomy through yeast fermentation of our feedstock, we will explore the use of the engineered starch-degrading strains to produce industrially relevant molecules. Examples of these molecules include the building blocks adipic acid and polyhydroxyalkanoates (market size of ~5 billion and ~80 million USD, respectively) as well as amino acids that could be used for human and animal feed (market size of ~30 billion USD).

578d Chemical and Structural Characterization of Lactose-Derived Novel Galacto-Oligosaccharides – an *in Vitro* and *In Vivo* Prebiotic Activity Assessment

Sonali Mohapatra, Meijun Zeng, Jee-Hwan Oh, Jan-Peter van Pijkeren, Xuejun Pan

The present study focused on the development of a facile and high-yield process for converting lactose into prebiotic galacto-oligosaccharides (GOS) with health-promoting properties. Lactose was used as a substrate to synthesize GOS using concentrated sulfuric acid (64–84% H_2SO_4) as reaction medium via an acid-catalyzed glycosylation reaction. An optimal GOS yield of 96% with minimal residual side products (such as galactose, glucose, anhydro sugars, and sugar degradation products) was obtained by using an initial lactose loading of 50% w/w in 76% H_2SO_4 after reaction at 70 °C for 20 minutes. Characterization of GOS revealed the presence of di-, tri-, and tetra-saccharides (47%, 25%, and 16%, respectively) and a small portion of larger oligosaccharides (DP up to 6). *In vitro* fermentation studies demonstrated that GOS promoted the growth of selected Lactobacillus strains, such as *Bifidobacterium bifidum* -ATCC 29521 and *Lactobacillus brevis* -ATCC 8287, comparable to commercial GOS. *In vivo* experiments in 8-week-old mice models confirmed an 8- and 3-fold higher persistence of *L. brevis* with supplementation of synthesized GOS at the 4th and 8th day, respectively, post termination of the probiotic in the feed system. Future studies on the persistence of *L. brevis* with a combination of GOS and cows and goat milk individually can help advance our knowledge of the effect of oligosaccharides supplemented with different milk on the gut environment.

578e Effects of Feedstock Preprocessing on the Wood Derived Organic Acid Yields

Matthew Luce, Sampath Karunarathne, Ravikant Patil, Peter van Walsum, Stephanie Ossai, Eliezer A. Reyes Molina, C. Luke Williams, Sampath Gunukula, Clayton Wheeler

Levulinic acid is a platform chemical, and it can be upgraded to a range of chemicals and fuels. Levulinic acid can be derived from lignocellulosic feedstock using the acid hydrolysis and dehydration process. The effective preprocessing of lignocellulosic feedstock is needed prior to its conversion to levulinic acid at commercial scale to address the flowability and downstream processing challenges. However, the studies to understand the effects of lignocellulosic preprocessing on the levulinic acid yields, by-product formation, and char characteristics have been limited to non-existent. In this research, we use a bench scale reactor to study the effects of preprocessed forest biomass using the chemical preconditioning system, steam pretreatment method, and thermal pretreatment method. Further, the effect of increase in the concentration of levulinic acid in the reaction mixture on the kinetics of acid hydrolysis and dehydration and the yields of levulinic acid will be determined. The initial results indicate that the removal of hemicelluloses of pine wood chips using the chemical preconditioning system increase the levulinic acid yields in comparison to the yields of levulinic acid produced from untreated whole pine wood chips. Further, the results show that chemical preconditioning of wet pine wood chips at 170 °C and a residence time of 30 minutes led to high yields of levulinic acid. Finally, the organic acid yield results indicate low initial moisture contents of feedstock gave lower yields but that air drying to 25% moisture content had no significant impact and that a changing particle size between 1 mm to ~18 mm does not have significant impact on the organic acid yields.

578f Catalytic Membrane Reactor for Conversion of Biomass Feedstocks to Levulinic Acid

Nhiem Cao, Xianghong Qian, Ranil Wickramasinghe

Global warming due to release of greenhouse gases coupled with increasing world energy demands requires immediate development of renewable resources for production of fuels and chemicals. Though many technologies exist to convert biomass feedstocks to fuels and value-added products, commercialization has been slow. The high cost of production of biobased chemicals and fuels frequently makes them uncompetitive compared to fossil fuel derived products. Here a patented catalytic membrane reactor has been developed that can convert a variety of lignocellulosic biomass feed stocks (corn stover, corn fiber, wheat straw, almond and walnut shells, coffee ground, rice husks etc.) to levulinic acid with high yield and high selectivity. Levulinic acid is considered a critical platform chemical for the production of numerous chemicals, resins, fibers etc as well as the intermediate to aviation fuels. Today levulinic acid is largely produced from fossil fuels. The levulinic acid market is valued at over $27 billion annually.

Our catalytic membrane converter has a ceramic membrane substrate immobilized with a dual functional catalyst by grafting polystyrene sulfonic acid (PSSA) and polyvinyl imidazolium chloride (polyionic liquid, PIL) chains from the surface of the membrane. The grafting is shown schematically in Figure 1. The PSSA chains catalyze biomass hydrolysis and the subsequent conversion to levulinic acid. The neighbouring (PIL) chains help solubilize lignocellulosic biomass and enhance the catalytic activity of the PSSA chains. The PSSA chains were synthesized *via* surface-initiated atom-transfer radical polymerization whereas the adjacent PIL chains were synthesized *via* UV-initiated free radical polymerization.

Our results indicate up to 80% of levulinic acid can be obtained from almond hulls without any pretreatment. Moreover, over 90% yield can be reached from almond shells.

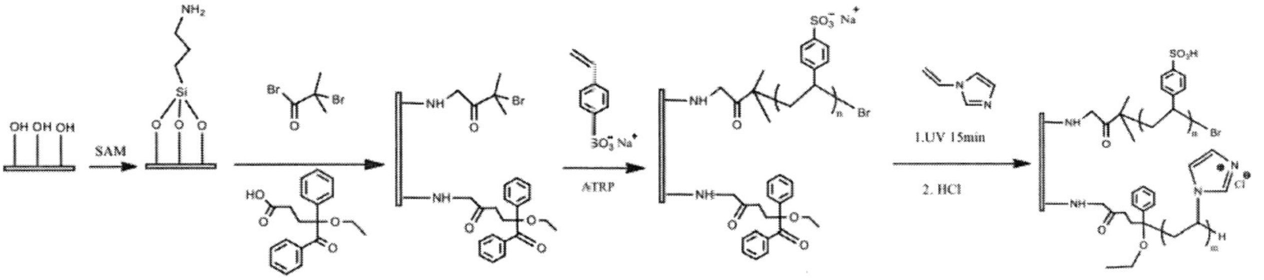

578g Dynamic Effect of CO_2 Concentration during Biosuccinic Acid Fermentation Process: Model Development on Glucose and Sugars-Rich Industrial Waste.

Francesco Vigato, Irini Angelidaki, John Woodley, Merlin Alvarado-Morales

Biosuccinic acid is a C4 molecule with a core role in the green transition of the biochemical industry[1]. Its fermentation process demonstrated several advantages compared to the petrochemical route where one of the key benefits of this biotechnology consists in the microbial uptake of CO_2 and its fixation in the final product: biosuccinic acid. The positive effect of CO_2 in the cells is largely proved: an increase in its availability corresponds to an increase in the production of biosuccinic acid[2,3]; a decrease to favor the competitive metabolic pathway. Even if this metabolic shift is well known, the consumption of CO_2 is not yet considered in most of the models of the research. The recent approach of describing the specific cell growth rate using a double substrate system, as in the case of the Monod equation, is a valid alternative but not sufficient to describe the dynamics in the carbon flux based on the CO_2 concentration[4]. As a result, a new and innovative structured mechanistic model was developed to concatenate the consumption of CO_2 in biosuccinic acid production.

In total, seven state variables are predicted from the model simulation: biomass (g/L), sugar (g/L), biosuccinic acid (g/L), acetic acid (g/L), formic acid (g/L), and CO_2 in the liquid (g/L) and gaseous phase (kPa). From the kinetic equations, Monod single substrate determines the specific microbial growth (μ), and accordingly the production of byproducts. On the opposite, the formation of biosuccinic acid is dynamically defined by the specific production rate (β) and strictly link with the concentration of the two relevant substrates: CO_2 and sugar. Accordingly, when CO_2 reaches the stress limit concentration the production of biosuccinic acid decreases without interfering with biomass and byproducts formation.

A provisional calibration with literature data was performed to test model outputs and simulation behavior. From the identifiability test and local sensitivity analysis, the parameters with less influence on the state variables were neglected from the subsequent estimations. The first parameter estimation with preliminary experimental results was conducted by the minimization of the errors between model output and fermentation results. Of particular interest was the model capacity to predict the shift in pathway when the CO_2 concentration was reduced below 20% of the saturation point by a significant reduction in biosuccinic acid yield. After, the developed model was utilized as a driver to conduct CSTR tests with suspended cells. In this series of experiments, pure glucose and medium were fed on the bioreactor inlet and the tank sparged with a constant flow of CO_2. Later, the pure glucose solution was replaced with a sugars-rich industrial waste and the CO_2 with a gas mock mixture of CO_2/CH_4 to emulate the biogas composition. In a second set of experiments, the model demonstrated a high accuracy in respects to the experimental results (std $\pm$ 5%). A consequent optimization analysis was implemented on titer and productivity. A maximum of 16 g/L and 1.4 g/L·h were achieved in the continuous fermentation.

Overall, the prediction of the dissolved CO_2 profile can give useful information on microbial consumption and uptake. Its simulation is a tool to highlight the potential of carbon fixation of the process and the use of biogas an example for the cleaning of similar industrial off-gasses rich in CO_2.

On these bases, the fermentation process moves from a stand-alone technology to an integrated system with the production of a high value-added product and the consumption of CO_2 from raw or waste organic and inorganic carbon sources. Last, the identification of the limited CO_2 concentration has a fundamental role to optimize the whole process outcome. Setting the operational condition to avoid the shift in the metabolic pathway will benefit in a positive way the production of biosuccinic acid and, indirectly, the downstream cost related to the less presence of byproducts compare to the principal target.

References

(1) E4tech; RE-CORD; WUR, and. *From the Sugar Platform to Biofuels and Biochemicals*; Final report for the European Commission, 2015; p 183. https://ec.europa.eu/energy/sites/ener/files/documents/EC Sugar Platform final report.pdf.

(2) Gunnarsson, I. B.; Alvarado-Morales, M.; Angelidaki, I. Utilization of CO2 Fixating Bacterium Actinobacillus Succinogenes 130Z for Simultaneous Biogas Upgrading and Biosuccinic Acid Production. *Environmental Science and Technology* **2014**, *48* (20), 12464–12468. https://doi.org/10.1021/es504000h.

(3) Vigato, F.; Angelidaki, I.; Woodley, J. M.; Alvarado-Morales, M. Dissolved CO2 Profile in Bio-Succinic Acid Production from Sugars-Rich Industrial Waste. *Biochemical Engineering Journal* **2022**, *187*, 108602. https://doi.org/10.1016/j.bej.2022.108602.

(4) Rigaki, A.; Webb, C.; Theodoropoulos, C. Double Substrate Limitation Model for the Bio-Based Production of Succinic Acid from Glycerol. *Biochemical Engineering Journal* **2020**, *153* (September 2019), 107391. https://doi.org/10.1016/j.bej.2019.107391.

578h Platform Chemicals from Hardwood Black Liquor Via Hydrothermal Liquefaction: Influence of Process Conditions on Product Yields and Quality

Harisankar S, Vinu Ravikrishnan

Black liquor is a complex aqueous by-product from the paper and pulp industry comprising of lignin residues and inorganic chemicals such as NaOH and Na_2S. Valorization of black liquor through hydrothermal liquefaction (HTL) is a sustainable alternative to the current practice of burning the black liquor in boilers to generate bioproducts and biochemicals. In this work, HTL of hardwood black liquor from Eucalyptus is carried out, and the influence of process variables such as temperature (250 – 400 °C), biomass loading (7.5 – 30 wt.%), and residence time (15 – 60 min) on yield and quality of the products is studied. Out of the various operating parameters, temperature and solid loading governed the yield of different product phases, while only temperature had a major role on the product quality.Increasing the operating temperature from 250 °C to 400 °C decreased the yield of liquid product from 79.8 wt.% to 52.3 wt.%, while increasing the residence time from 15 min to 60 min barely changed the yields by ±5 wt% without any prominent effect on the product yields. Increasing the solid loading from 7.5 wt.% to 30 wt.% increased the yield of hydrochar from 21.9 wt.% to 48.6 wt.%. Phenols, aliphatic hydrocarbons and ketones were the major organic compounds detected in the liquid product. The lignin structure depolymerizes at 250 °C to form methoxy phenols, which gets then hydrolysed at temperatures above 300 °C to give dihydroxybenzenes such as catechol and its derivatives. The dihydroxybenzenes further undergo dehydration and dealkylation at subcritical conditions to produce phenol and alkyl phenols. Increasing the operating temperature led to an increase in the selectivities to phenol, cresols and xylenols. Maximum selectivity to phenol (23.7%) was obtained at 350 °C, 30 min and 15 wt.% solid loading. Na^+ and K^+ were the abundant ions detected in the liquid phase. Increasing the reaction temperature led to an accumulation of water-soluble salt ions in the liquid product, while the ion concentration in hydrochar decreased with temperature. The optimum operating conditions for maximum selectivity towards phenol in liquid product is shown to be 350 °C, 45 min and 30 wt.% solid loading. The optimum temperature for maximum energy recovery from hydrochar is 400 °C, 30 min, 15 wt.% solids. The findings suggest that HTL is a promising treatment method to recover energy through hydrochars as well as produce platform chemicals using waste black liquor from the pulp mills.

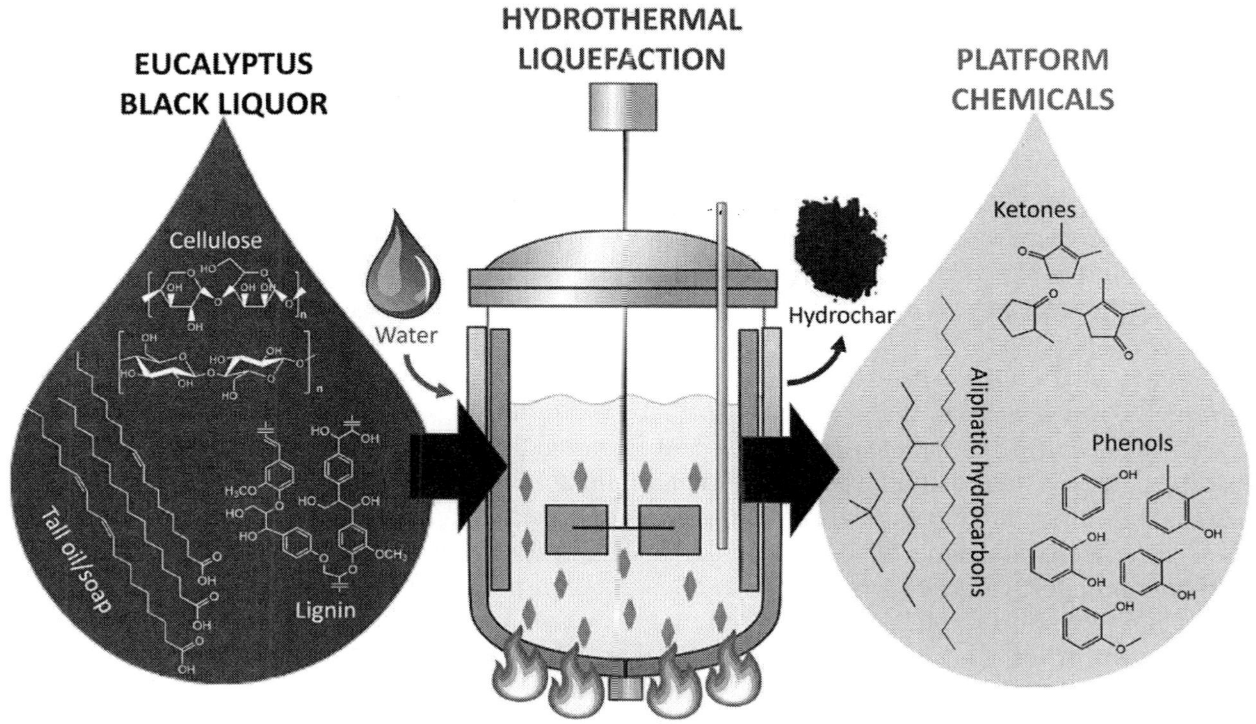
EUCALYPTUS BLACK LIQUOR
HYDROTHERMAL LIQUEFACTION
PLATFORM CHEMICALS
Cellulose
Tall oil/soap
Lignin
Water
Hydrochar
Ketones
Aliphatic hydrocarbons
Phenols

549a Reaction Kinetics and Physical Properties of Carboxymethyl Cellulose-Based Hydrogel Blends

Samudra Gupta, Javen Weston

Carboxymethyl cellulose (CMC) is a commonly studied polymer made by chemically modifying cellulose to increase water dispersability. CMC is widely used in hydrogel applications because of its biocompatibility and biodegradability. CMC hydrogels are used as absorbents, for food packaging, as additives in pharmaceutical formulations, and for other applications. CMC hydrogels can be synthesized using a variety of chemical and physical crosslinking agents. Here, ethylene glycol diglycidyl ether (EGDE) is used as a chemical crosslinker. The optimal cross-linking temperature was determined using *in-situ* oscillatory shear rheology, and other properties, such as the water vapor permeability, swelling ratio, tensile strength, and % elongation at break of the hydrogels were also measured to assess their usefulness for different applications. These hydrogels were found to have unsatisfactory bulk and mechanical properties for food packaging materials, so blends of CMC with various polysaccharides have been investigated to search for synergy between the different biopolymers. In particular, κ-Carrageenan and Xanthan gum were chosen for this study, both are common anionic polysaccharides with excellent gelling, stabilizing ability, emulsifying, and thickening properties. These blends were chosen due to CMC being is easy to use, readily available, and cheap. Additionally, this study focuses on finding a correlation between rheology, and bulk properties, such as tensile strength, water vapor permeability, and % elongation at break.

549b Cellulose Nanofiber-Alginate Biotemplated Composite Co_3O_4 Aerogels for Pseudocapacitor Electrodes

Galen Mandes, Vani Verma, Anthony Presot, Claire Tsay, Felita Zhang, Paul Trackey, Rosemary L. Calabro, Enoch Nagelli, Joshua A. Maurer, Stephen F. Bartolucci, F. John Burpo

Supercapacitors are a critical technology for energy applications that require rapid charge/discharge cycles. Electrochemical pseudocapacitors use metal oxides that undergo redox reactions to generate additional charge storage. A factor that regulates charge storage is the surface area of the capacitor, which can be greatly improved per unit mass by implementing a porous structure. While porous metal oxide pseudocapacitors have been fabricated previously, the mechanical durability of the materials has not been adequately demonstrated. A porous, mechanically strong, composite material with tunable pore size, nanowire length, diameter, and material phase offers the possibility of electro-mechanical materials that enable device fabrication at multiple length scales with an overall decrease of systems mass and corresponding increase in performance metrics. Previous studies demonstrate the use of biopolymer hydrogels to biotemplate composite metal nanowire porous electrodes.[1–3] In this work, we demonstrate a Co_3O_4 biotemplated pseudocapacitor synthesized using a carboxymethyl cellulose nanofiber (CNF)-alginate composite hydrogel scaffold. The hydrogels are equilibrated in $CaCl_2$ / $CoCl_2$ salt solutions and chemically reduced with $NaBH_4$, where the ratio of cations determines the resulting degree of ionic crosslinking and transition metal phase. After supercritical drying, aerogel pyrolysis and thermal oxidation enhance the composite material conductivity and achieve electrochemically active Co_3O_4. Material characterization is demonstrated with scanning electron microscopy, x-ray diffractometry, thermal gravimetric analysis, electrochemical impedance spectroscopy, cyclic voltammetry, and compressive loading. The resulting free-standing electrodes provide rapid charge and discharge and a mechanically robust material, providing potential applications as both a structural and energy storage material for a wide range of military and commercial applications demanding light-weight power and energy materials.

[1] F.J. Burpo, M.Y. Ryu, E.A. Nagelli, E. Onuomadonkeng, Gelatin-Cellulose Nanofiber Biotemplated Platinum Nanowire Porous Fibers, (2019).

[2] J.F. Burpo, N.A. Mitropoulos, A.E. Nagelli, L.J. Palmer, A.L. Morris, Y.M. Ryu, K.J. Wickiser, Cellulose Nanofiber Biotemplated Palladium Composite Aerogels, 23 (2018). https://doi.org/10.3390/molecules23061405.

[3] J.F. Ohmura, F.J. Burpo, C.J. Lescott, A. Ransil, Y. Yoon, W.C. Records, A.M. Belcher, Highly adjustable 3D nano-architectures and chemistries via assembled 1D biological templates, Nanoscale. 11 (2019) 1091–1102. https://doi.org/10.1039/C8NR04864A.

549c Aptes-Molecularly Imprinted Polymer Modified Cellulose Nanocrystals for Carbofuran Detection

Sadat Kamal Amit, Diego Gomez-Maldonado, Tiana Bish, Maria Soledad Peresin, Virginia Davis

This talk describes a novel modification of cellulose nanocrystals (CNCs) to enable their use as selective sensors for detecting emerging contaminants in water. Carbofuran, a broad-spectrum insecticide and nematicide, was used as a model analyte. First, the CNCs were modified with 3-aminopropyl-triethoxy silane (APTES) to impart hydrolytic stability. A molecularly imprinted polymer (MIP) was developed and applied as a coating to CNC-APTES film to enable selective detection. The mechanical properties and hydrolytic stability of pure CNC and modified free-standing films were investigated to explore their potential in device fabrication and applications requiring immersion in water. CNC films did not retain their integrity in water. The APTES modification enhanced the films' hydrolytic stability without altering tensile strength. The MIP layer reduced the tensile strength of the films but also provided hydrolytic stability. The sensitivity and selectivity of CNC-APTES-MIP were investigated using quartz crystal microbalance with dissipation (QCMD). The material exhibited selective sensing of carbofuran compared to the herbicide 2,4-dichlorophenoxyacetic acid (2,4D), which has a similar benzene-based structure. The sensitivity was as low as 0.03 ppm, lower than standard commercial sensors. These results highlight CNCs' potential for sensing a wide range of analytes and their potential use in multiplexed devices for the simultaneous detection of multiple analytes.

549d Functionalized Cellulose Nanocrystals in Agriculture: Synthesis and Characterization for Use As a Plant-Based Nanocarrier of Active Biomolecules

Delaney Clouse, Elise Collins, Mariya Khodakovskaya, Virginia Davis

This investigation describes the development of novel bioactive molecule-conjugated cellulose nanocrystals for use as a targeted nanocarriers in agricultural applications. The United Nations estimates that agricultural food production must increase by 60% by 2050 to feed a growing world population. Herbicides and pesticides are critical to agricultural yield, but current delivery methods result in as little as 10% of the agrochemical reaching the plant with the remainder being lost to the environment. This can result in negative environmental consequences such as air, groundwater, and soil contamination. Nanocarriers are promising for increasing the efficiency of agrochemical application and reducing the potential adverse environmental effects of their use. Cellulose nanocrystals (CNCs) provide the opportunity to utilize a forestry and agricultural waste product to enhance agricultural yields. These nanocrystals have the small size for transport into plants, and surface functional groups to react with desired agricultural chemicals while maintaining crop biocompatibility and minimal environmental toxicity. This work describes the novel surface functionalization of cellulose nanocrystals with model agricultural biomarkers. Understanding biomolecule-cellulose-plant interactions is critical to the future development of agricultural bionanotechnology.

Commercially supplied sulfuric acid hydrolyzed cellulose nanocrystals were reacted with model agricultural biomolecules, such as 2,4-dichlorophenoxyacetic acid (2,4-D), using novel combinations of aqueous chemistries. The resulting functionalized CNCs were characterized using, spectroscopy, thermal analysis, elemental analysis, and microscopy techniques. The degree of substitution was calculated based on the number of available surface hydroxyl groups. Results confirmed the agricultural agents' attachment to the cellulose nanocrystals' surface. Future collaborative investigations with plant biologists will work to understand the effect and viability of cellulose nanocrystals as a nanocarrier in delivering biomolecules to plants. This investigation lays the framework for a future cyclic agricultural system, where cellulose nanocrystals can be extracted from crop waste, reacted with biomolecules of interest, and then directly delivered to plants, thus creating a more targeted effect while simultaneously reducing waste and increasing crop yield.

549e The Modification of Nanocellulose with Atomic Precise Metal Nanoclusters

Zhaoxian Qin, Yue Yuan, Zhaohui Tong

New functional materials are of great importance for the design of highly sensitive and high-precision sensors for chemicals and hazardous substances. The selective, precise, and swift measurement of chemical substances is critical in various areas such as environmental protection, the food industry, occupational safety, and forensics. Recently, the burgeoning metal nanoclusters (MNCs) with well-resolved crystal structures show great prospects in fundamental research and various applications including luminescence, medicine, catalysis, energy, and biology owing to their unique physiochemical properties. Additionally, with the ever-growing interest in sustainable materials, nanoclusters are assembled with renewable nano cellulose together to serve as color-changing ink, sensors for heavy metals, and antibacterial materials based on the intense fluorescence and biocompatibility of MNCs. However, it remains unclear which MNC in the mixture plays a major role in the final properties of nanomaterials since the MNCs used in reported MNCs/nanocellulose materials are mixtures of MNCs, not monodispersed and atomic precise MNCs. And the understanding of the correlation between MNCs' structure and properties is limited till now for the lack of suitable nano-assembling models.

In this study, a series of stable and high fluorescence homo- and heterogeneous nanoclusters were prepared and characterized by an array of advanced instruments including UV-vis spectroscopy, ESI-MS, XPS, and single-crystal X-ray diffraction. These nanoclusters were further deposited on the uniform nanocellulose crystals (NCs) to form sensors to detect heavy metal ions in water. Interestingly, the quantity yield of MNCs has been significantly improved in the presence of NCs, which suggests that NCs could improve the sensitivity of the sensor to metal ions' detection in water. Moreover, the dopant of Cu species in MNCs has been proven to be the key to the high stability and high sensitivity of MNCs-NCs nanocomposites in the nanosensor design. Besides, the pure silver cluster deposited on NCs is not only sensitive to metal ions in water but also shows an excellent anti-bacterial property. This work provides a good example of the design and application of atomic precise MNCs-NCs nanocomposites in the field of highly-sensitive sensors.

549f Cellulose Nanocrystals Improve Pathogen Resistance in Plants

Henry Squire, Autumn Lee, Natalie S. Goh, Jeffrey Wang, Cerise Wong, Markita Landry

Plants are capable of mounting an immune defense in response to infection, wounding, or environmental stress. The immune response of a plant is often mediated through recognition of conserved molecular motifs associated with stress. Peptides of bacterial origin, fungal oligosaccharides, and small molecule hormones for example can all elicit an immune response in plants. Treating plants with these molecular motifs prior to exposure to stress, termed immune priming, can improve plant resilience to the subsequent stressors. Here we demonstrate use of cellulose nanocrystals (CNCs) as an immune priming tool for *Arabidopsis thaliana* combining phenotypic and transcriptomic data to validate the efficacy of CNCs as an immune primer and probe underlying biological mechanisms. First, we show that pretreatment of *Arabidopsis* with CNCs significantly reduces infection of plants by the pathogen *Pseudomonas syringae* in a dose dependent manner. Next, RNA sequencing and RT-qPCR data suggest *Arabidopsis* recognizes CNCs as cell-wall damage, potentially explaining the observed immune response. Further physiological characterization of plant cell wall reorganization in response to CNCs further supports this hypothesis. This work provides a foundation for future investigation of cellulose-based materials as immune primers, particularly in crop species. Given their ease of synthesis, low cost, and bio/environmental compatibility, cellulose-based materials could serve as an alternative pathogen control methodology for agricultural scale application.

549g Whole Biomass Aerogel-Based Triboelectric Nanogenerator for Energy Harvesting and Self-Powered Sensing

Ruolin Wang, Zhenhui Jin, Haishun Du, Jiansong Chen, Longwen Li, Yi-Cheng Wang, Xuejun Pan

Triboelectric nanogenerator (TENG) is an emerging ambient technique that converts mechanical energy into electrical energy, demonstrating promising capabilities as distributed micro-power supplies, self-powered sensors, and more. However, conventional materials used in TENGs are mainly synthetic polymers, such as polytetrafluoroethylene (PTFE) and polyvinylidene fluoride (PVDF), which are neither sustainable nor biodegradable. Many efforts have been made to fabricate sustainable TENGs with natural polymers, such as cellulose, chitosan, and alginate. In a previous study, we demonstrated a novel method for fabricating cellulose II aerogels using lithium bromide trihydrate (LBTH) as a solvent. The obtained aerogels were homogeneous and highly flexible with high porosity and large specific surface area, and the TENG made from the aerogels exhibited excellent performance in mechanical response sensitivity and electrical output. However, the preparation of pure cellulose from biomass requires complicated pulping processes that involve the utilization of harmful chemicals and additional energy and water consumption. Therefore, it is promising to fabricate aerogels directly from whole lignocellulosic biomass. In this study, we developed a method to fabricate lignocellulosic biomass aerogels directly from whole poplar biomass through the dissolution-regeneration process in LBTH. We systematically investigated and quantified the effect of different parameters, including aerogel preparation conditions, chemical composition, and surface modification, on the electric output performance. Moreover, the fabricated TENG can serve as a self-powered pressure sensor to sense different external stimuli under different working modes. Overall, this study showcases a sustainable and economically feasible fabrication method of whole lignocellulosic biomass aerogels and demonstrates their capability for constructing TENGs and self-powered sensors, offering new ideas in developing more sustainable and environmentally friendly energy harvesting and self-powered sensing devices.

549i Production and Characterization of Holocellulose Micro and Nano Fibers in Alternative Solvents

Samarthya Bhagia, Donna Johnson, Katie Copenhaver, Soydan Ozcan

Cellulose-hemicellulose (holocellulose) fibers of micron to nano scale dimensions can be used to make a wide variety of renewable materials like holocellulose reinforced composites and substrates for flexible electronics. However, the cost of production needs to be lowered for commercialization by reducing the energy consumption in milling and drying processes. Therefore, a variety of new water-based solvents were utilized to see their impact on pulp fibrillation performance, fiber morphology and chemical structure. These aqueous solvents were made by adding low concentrations of additives like carboxymethylcellulose that acts as a viscosity modifier, or sodium hydroxide-urea which was found to be a top ranking additive from molecular dynamic simulations. Holocellulose micro/nano fibers (CNF) fibrillated in these alternative solvents were studied for energy consumption, size distribution, morphology, thin film forming qualities and composition. In particular, ^{13}C solid-state NMR based techniques were investigated to understand the changes to the crystallinity of cellulose fraction in native and never-dried state and without hemicellulose removal by wet chemistry.

503a Fractionation of Lignocellulose and Tandem Preservation of Lignin and Carbohydrates Using Aldehyde Sulfonic Acid

Danni Xie, Qiang Yang, Yunxuan Wang, Chang Geun Yoo

The effective fractionation of lignocellulose with low energy inputs while preserving the obtained cellulose, hemicellulose, and lignin is the key to utilize lignocellulose economically and effectively. This study presents an aldehyde sulfonic acid enabled mild-condition (e.g., 60 °C for 2 h or 80 °C for 15 min) process that can near-completely fractionate hardwood chips into water-soluble lignin, xylose-derived dioxolo, and cellulose. The bifunctional aldehyde sulfonic acid can depolymerize hemicellulose and lignin by the sulfonic acid group and tandemly react and preserve the obtained xylose and the depolymerized lignin through the aldehyde group autocatalyzed by the sulfonic acid group. The obtained lignin is enriched with beta-O-4 linkage, and is not condensed, and is water-soluble. The water-soluble lignin can be separated as cleavable (cyclic acetal-linked) surfactant through membrane dialysis. Xylose can be effectively recovered from xylose-derived dioxolo through acidolysis. The obtained cellulose can be directly hydrolyzed to glucose using cellulolytic enzymes. The unreacted aldehyde sulfonic acid can be recycled for reuse.

503b Top-Down Synthesis of Potential Prebiotic Gluco- and Xylo-Oligosaccharides from Corn Stover Via Simultaneous Hydrolysis and Glycosylation in Concentrated Sulfuric Acid

Meijun Zeng, Sonali Mohapatra, Jeehwan Oh, Jan Peter van Pijkeren, Xuejun Pan

Valorization of lignocellulosic biomass residues into functional oligosaccharides with prebiotic activity has attracted great interest from both environmental and economic perspectives. Cellulose and hemicelluloses, the two structural polysaccharides naturally present in the matrix of lignocellulosic biomass, can be depolymerized into soluble oligosaccharides, such as cello-oligosaccharides and xylo-oligosaccharides (XOS), which are non-digestible and fermentable by the health-promoting gut microbiota, making them potential prebiotics. Currently, XOS are industrially produced from corncob xylan via alkaline extraction followed by enzymatic hydrolysis by xylanases. However, cellulose (~30% in corncob) remains unutilized, and the process economy is limited by the high cost associated with enzymes and the pretreatment step. Herein, we report a facile and efficient chemical method (top-down synthesis) for the co-production of potential prebiotic gluco-oligosaccharides (GlcOS) and XOS mixtures from unpretreated corn stover. Specifically, cellulose and xylan in corn stover are subject to simultaneous hydrolysis and glycosylation (SHG) in concentrated sulfuric acid (60–84%) under mild conditions (50–70 °C, 5–60 min). After the removal of lignin residues, a soluble oligosaccharide product is obtained, which is identified as a mixture of GlcOS and XOS with mostly 2–6 monomer units, linked by diverse glycosidic linkages (α- and β-(1→6/4/3/2/1)). The total yield of corn stover-derived oligosaccharides (CS-OS) is close to the theoretical yield, with minimal loss into monosugars and acid-induced degradation products (e.g., furfural, 5-hydroxymethylfurfural). The prebiotic activity of CS-OS was assessed via in vitro fermentation with different lactic acid bacteria. CS-OS could promote the growth of certain Lactobacillus and Pediococcus strains (potential probiotics), outperforming two commercial prebiotic oligosaccharides (fructo- and isomalto-oligosaccharides). Furthermore, a mechanism study was conducted to reveal the impact of oligosaccharide size (degree of polymerization, DP) and glycosidic linkages on their utilization by probiotic bacteria. Preferrable utilization of DP 2–3 over DP 4–7 oligosaccharides was observed, and the bacterial ability to utilize α/β-(1→6/4) linkages varies significantly depending on the available genomes essential for oligosaccharide transport and catabolism. Overall, this study demonstrated the production of value-added oligosaccharides as potential prebiotics from lignocellulosic biomass, which is an emerging sector in response to the burgeoning market of prebiotics.

503c Alkaline Pretreatment to Fractionate Industrial Hemp for Biochemical Conversion and the Effects on Polysaccharide and Lignin Recovery

Ryan Stoklosa, Renee J. Latona, Charles A. Mullen, Bryan Berger, William Fahy, Michael P. Timko

The passage of the 2018 Farm Bill in the United States directed the USDA to establish national regulations for the cultivation of industrial hemp in the U.S. This has allowed agricultural producers to start growing industrial hemp to be used in new application routes as a feedstock for bio-based fuels and chemicals. The previous restrictions on planting and harvesting hemp have not allowed this feedstock to be sufficiently researched in the context of biochemical conversion compared to other lignocellulosic crops such as woody biomass or perennial grasses. Thus, the aim of this research is to provide a more thorough understanding on the deconstruction of industrial hemp to access both polysaccharide and lignin fractions. In this work industrial hemp was pretreated using alkaline chemistry to determine the effect on both hemicellulose and lignin removal. Increasing chemical severity confirmed that cellulose content was maintained after pretreatment, but hemicellulose and lignin removal increased under the same conditions. Enzymatic hydrolysis of the pretreated hemp produced high yields of sugar recovery (i.e. glucose and xylose) to generate a hydrolysate media for downstream fermentation. The resulting pretreatment liquor was utilized as a source for lignin separation and recovery. The recovered hemp lignin from pretreatment underwent chemical characterization using microscale analytical pyrolysis to characterize constituent monomeric components.

503d High Solids Loading Processing of Hydrothermally Pretreated *Miscanthus* x *Giganteus* for Production of Anthocyanins, Sugars and Lignin-Rich Residue

Shivali Banerjee, Bruce S. Dien, Erik J Sacks, Vijay Singh

Abstract

Miscanthus x giganteus is a widely recognized bioenergy crop that finds potential applications in a biorefinery for the production of a wide range of biofuels and biochemicals (Rivas et al., 2019). In this study we focused upon the lab-scale and pilot-scale bioprocessing of miscanthus at high solids loading for recovery of value-added products such as anthocyanins, microbial lipids and lignin-rich residue. Hydrothermal pretreatment was studied for recovering anthocyanins. The optimized pretreatment conditions (170 °C, 10 min) led to a recovery of 94.3±1.5% w/w of the total anthocyanin content and also improved the enzymatic digestibility of the biomass, allowing recovery of 70.6±0.5% w/w of glucose at the end of 72 h of enzymatic saccharification. Based upon the lab-scale studies, the pretreatment step was scaled-up with 50 kg of miscanthus with 50% solids loading using a continuous pretreatment reactor (Cheng, Dien, Lee, & Singh, 2019) that led to a recovery of 70% w/w of total anthocyanins present in the biomass. The pretreated biomass was subjected to enzymatic saccharification at 10-50% solids loading using fed-batch operations. The highest sugar yields were obtained for 50% solids loading. The sugar concentrations of the hydrolysate were 140 g/L and 80 g/L for glucose and xylose, respectively. The hydrolysate was further used for the production of microbial lipids. While the residue left after enzymatic saccharification was found to be enriched in lignin that could further be diverted toward the production of aviation fuel precursors. Hence a detailed biorefinery concept was demonstrated by bioprocessing of miscanthus leading to the generation of multiple products.

References

Cheng, M.-H., Dien, B. S., Lee, D. K., & Singh, V. (2019). Sugar production from bioenergy sorghum by using pilot scale continuous hydrothermal pretreatment combined with disk refining. *Bioresource Technology, 289*, 121663. doi:https://doi.org/10.1016/j.biortech.2019.121663

Rivas, S., Vila, C., Alonso, J. L., Santos, V., Parajó, J. C., & Leahy, J. J. (2019). Biorefinery processes for the valorization of Miscanthus polysaccharides: from constituent sugars to platform chemicals. *Industrial Crops and Products, 134*, 309-317. doi:https://doi.org/10.1016/j.indcrop.2019.04.005

503e Production of Low-Cost and Highly Fermentable Sugar from Corn Stover Via Chemical-Recovery-Free Deacetylation and Mechanical Refining (CRF-DMR) Process

Xiaowen Chen, Junyong Zhu, Bin Yang, Joshua Yuan, Eric Tan, Ryan Davis, Nancy Dowe, Zhihua Jiang, Yudong Li, Jorge Arreola Vargas, Fernando Roberto Paz Cedeno, Carlos Driemeier, Michael E. Himmel

The pioneer 2nd generation cellulosic ethanol plants faced challenges due to the low fermentability of sugar hydrolysates produced by dilute acid pretreatment. This technique depolymerizes hemicelluloses, mainly xylan, in corn stover biomass, improving the digestibility of pretreated biomass solids and resulting in a mixed glucose and xylose sugar hydrolysate through enzymatic hydrolysis. However, the high temperature and low pH conditions of the reaction cause rapid sugar degradation reactions, leading to the formation of furfural from xylose and 5-hydroxymethylfurfural (HMF) from glucose, both of which are strong fermentation inhibitors even at low concentrations. Additionally, hemicelluloses in corn stover are heavily esterified with acetyl groups, and during dilute acid pretreatment, these groups are released, forming acetic acid, another potent inhibitor of ethanol fermentation using common organisms like *Zymomonas* and *Saccharomyces cerevisiae*. The resulting sugar hydrolysate from dilute acid pretreated corn stover (hydrolyzed at 20% total solids) contains roughly 10 g/L of acetic acid, 5 g/L of furfural, and 0.5 g/L of HMF. As a result, only 50% of the xylose is utilized after 50 hours of fermentation when glucose has already been completely consumed, leading to an overall ethanol yield of around 80% after 75 hours of fermentation.

The National Renewable Energy Laboratory (NREL) has developed a solution to this issue in the form of a scalable, atmospheric pressure, and low severity biomass deconstruction and fractionation process known as the Deacetylation and Mechanical Refining (DMR) process. The DMR process has exhibited exceptional performance in converting corn stover to high-concentration, low-toxicity sugars at high yields. The process involves a dilute NaOH deacetylation step to remove the acetate in biomass at atmospheric pressure (90°C), followed by mechanical refining of the deacetylated biomass using a disk refiner and a Szego mill, both commonly used in the pulping and milling industry. The entire process is carried out at low temperatures without an acid catalyst, effectively preventing the formation of furfural and HMF. Additionally, the removal of acetyl groups significantly reduces acetic acid formation during sugar hydrolysate production, resulting in high fermentation efficiency. The hydrolysate produced by DMR from corn stover (hydrolyzed at 25% total solids) resulted in an ethanol yield exceeding 98% with xylose utilization exceeding 95% in only 40 hours of fermentation due to zero furan content and very low acetic acid concentration of less than 0.5 g/L.

The high sodium hydroxide usage in DMR process is beneficial to increase sugar yields while reducing enzyme usage. However, the cost of sodium hydroxide can be significant at $0.24/lb, and the high loadings required in the process can lead to high greenhouse gas (GHG) emissions. While NaOH recovery is common in the pulp and paper industry, it is complex and capital intensive, requiring an expensive recovery boiler, lime kiln, and recausticization cycle. These processes account for approximately 50% of the total capital investment for a 2000 tonne biomass/day biorefinery plant.

This study developed a chemical-recovery-free pretreatment technology for the DMR process, which potentially addresses current challenges and decrease the minimum sugar selling price. In this modified DMR process, NaOH is replaced by ammonium or potassium alkali and salts, which results in post-fermentation spent liquor that can be used as a fertilizer without requiring chemical recovery. In fact, ammonium or potassium sulfite pulping spent liquor (with low pH) has been used as a fertilizer in various countries. Additionally, the modified DMR spent liquor has a lower sulfur content due to the reduced sulfite charge. By partially absorbing CO_2 from downstream fermentation, the alkaline spent liquor becomes near pH-neutral, which is more beneficial for the environment and soil. This method not only reduces GHG emissions but also provides a carbon source for plants. By converting the pretreatment spent liquor into fertilizer without the need for wastewater treatment or chemical recovery, we could save at least $150MM in capital investment, and the revenue from the fertilizer could cover the cost of pretreatment chemicals. The return of nutrients, lignin, and biogenic CO_2 back into the soil would help balance nutrient loss and maintain soil organic carbon that is removed during the corn stover removal process. Furthermore, the chemical recovery-free process reduces the energy needed for the GHG-intensive calcination process, which could potentially reduce the carbon footprint of sugar production.

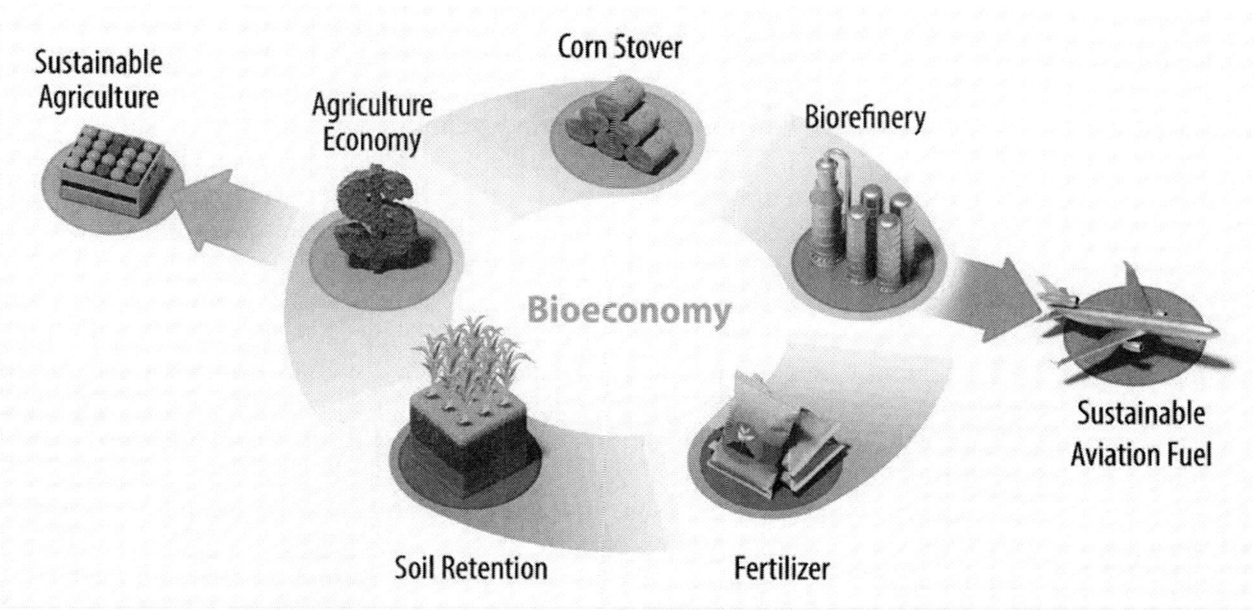

503f Switchgrass Deconstruction Mechanisms on Biomass Solubilization Elucidated through Characterization of Consolidated Bioprocessing Residues.

Rohit Kousika, Yunqiao Pu, Arthur Ragauskas, Yannick J. Bomble

The most significant obstacle to commercially viable biofuels and bioproducts remains the cost-effective breakdown of cellulosic biomass. Consolidated bioprocessing (CBP) has been acknowledged for its ability to decrease the number of process steps and eliminate the costs associated with additional saccharolytic enzymes. Current developments in the use of cellulolytic anaerobes like *Clostridium thermocellum* for deconstruction, which are highly successful at solubilizing recalcitrant biomass than industry-standard fungal cellulase, provide proof-of-concept for CBP. The solubilization of both herbaceous and woody non-pretreated biomass is significantly increased by the novel and promising CBP/CT using thermophilic anaerobes combined with cotreatment (i.e., milling during fermentation). In this study ball milling and disc milling operations were used along with anaerobic fermentation as part of the CBP/CT process. The residues from the milling followed by fermentation steps were analyzed using a variety of characterization techniques such as scanning electron microscopy (SEM), Heteronuclear single quantum coherence (HSQC) – Nuclear magnetic resonance (NMR), Gel permeation chromatography (GPC). Considerable differences in the microfibril structure were observed among the ball-milled and disc-milled switchgrass CBP/CT residues. The lignin, cellulose molecular weights, monolignol composition, and interunit linkages resulting from different milling conditions were investigated to understand the deconstruction mechanisms of CBP/CT. Moreover, the total lignin and sugar contents of the CBP/CT residuals provide an estimation of the process efficiency in depolymerizing the lignocellulosic matrix and fermentation of cellulosic and hemi-cellulosic sugars. Quantifying the effects of CBP/CT (mechanical disruption) on biomass will identify its current limitations and can help in developing potential improvement strategies along with the application of CBP/CT solid residuals for value-added products.

503g Corn Stover: Understanding the Impact of Anatomical Heterogeneity on Bioprocessing

William Otto, Dylan Cousins, David Hodge

Corn stover, the crop residue from harvesting field corn, is an abundant biomass resource that has been widely researched as a potential feedstock for bioproducts. Yet, it is unclear how the inherent heterogeneity present in this biomass impacts processing. To date, corn stover is treated as a homogenous feedstock. Anatomically, a corn stover plant consists of multiple parts that specialize in unique roles for plant survivability, each composed of different tissue type and structure. These differences pose challenges to bioprocessing such as handling and sugar recovery. Quantifying these differences will provide a fundamental understanding in corn stover biorefining that will result in new processing strategies to exploit anatomical heterogeneity for improved biorefining performance and better selectivity for targeted products. In this study, we identified six distinct anatomical components: stalk-rind, stalk-pith, cob, sheath, leaf, and husk. These components were manually separated, subjected to multiple pretreatment conditions, and characterized. Early results showed differences in chemical composition, bulk physical properties, and water sorption behavior, resulting in different pretreatment response and enzymatic hydrolysis with each fraction. This work explores the relationship between anatomical fraction and water sorption properties to understand processing limitations and enzyme accessibility. These results coupled with mass balance and techno-economic analysis will be instrumental to understanding the economic benefits and viability of biorefining preferentially rich, anatomical fractions.

536a Optimization of Biocrude Yield and Generated Wastewater Biodegradability in Hydrothermal Liquefaction of Corn Stover

Isamu Umeda, Meicen Liu, Yi Zheng, Chandan Mahata, Jiefu Wang, Zhiwu Wang, Jeawan Yoon, Sandeep Kumar

Hydrothermal liquefaction (HTL) can be one of the potential thermochemical approaches to produce biofuels and bioproducts because it does not require dewatering or drying and is able to use a wide variety of organic materials as feedstocks. In addition to biocrude and biochar, HTL also generates wastewater which may include toxic compounds. Therefore, we not only aim at the production of high quality cost-effective biocrude but also consider the HTL wastewater biodegradability. Biological treatment can be a less expensive option than other approaches, such as activated carbon and chemical treatment. The current research includes two objectives: 1) investigate the relationship among HTL operating conditions, the biocrude yields/properties and the characteristics of HTL wastewater to predict the wastewater biodegradability by using statistical analysis; and 2) optimize the biocrude yields/properties and HTL wastewater biodegradability.

The current study is focused on optimizing HTL of corn stover in the temperature range of 250 – 350 °C and reaction time of 1 – 60 min. The generated HTL wastewater will be used to cultivate three different fungal strains including *Aspergillus niger* strain NRRL 2001, *Phanerochaete chrysosporium* strain BKM-F-1767, and *Rhodotorula mucilaginosa* to evaluate the HTL wastewater biodegradability.

536c Solvent Mediated Extraction of Fast Pyrolysis Bio-Oils for Isolating Phenolics

Yaseen Elkasabi, Charles A. Mullen

Fast pyrolysis of agricultural residues produces a viscous liquid (bio-oil) that can serve as a precursor to renewable drop-in fuels and coproducts. While adaptation of bio-oils to a refinery serves as the ideal, the feasibility relies heavily on the oxygen content of said bio-oil. A relatively low oxygen content (< 16 wt% O) would lower both oil density and reactivity. Distillation of these bio-oils proceeds with significant yields of organic compounds. However, if oils cannot undergo distillation due to reactive instabilities, then separation methods are still desirable for further upgrading. A modified extraction method on whole oil could bypass distillation of unstable oils, which could also reduce heat duty and process costs if extraction is performed first.

In the past we've shown that one-ring phenolics can be extracted from distillates of partially deoxygenated bio-oils (<16 wt% O), thereby producing a hydrocarbon fraction in parallel. We will discuss the use of solvent-mediated extraction of whole bio-oils, (16 - 24 wt% O) based on toluene and isopropyl alcohol. Bio-oils of varying oxygen content and quality underwent extraction with toluene. The toluene soluble fraction (TS) underwent extraction with 10M NaOH, thus producing phenolics and hydrocarbon fractions. The Toluene insoluble fraction (TI) underwent extraction with isopropyl alcohol, producing soluble and insoluble fractions (IPA-S; IPA-I). We successfully isolated hydrocarbon fractions and one-ring phenolics without distillation, with the option to further upgrade the highly oxygenated IPA-S and IPA-I fractions. Specifically, the IPA-I fraction underwent py-GCMS with and without catalyst, which showed conversion to deoxygenated gasoline and jet fuel-range compounds.

536d Synthesis of Renewable Insecticides Possessing Tailored Functional Groups

Tejas Goculdas, Sunitha Sadula, Dionisios Vlachos

The carbamate insecticides market size is forecasted to reach a value of $341.6 million by the end of 2027 growing at a CAGR of 4.5%. This poses a significant environmental threat as commercial carbamate insecticides are produced from fossil fuel-derived phenols such as cresol. Synthesis of these insecticides involves strong acids and produces toxic waste streams. Bio-based insecticides are a more sustainable and often safer alternative for pest management in agricultural production. Prior art in this field is limited to testing bio-oil obtained from the pyrolysis of biomass and other mild fungicides derived from reductive catalytic fractionation of lignin. However, the insecticidal activity of these compounds is several orders of magnitude lower than that of commercial insecticides.

In this work, we implemented targeted synthesis to construct carbamate functional groups, the active component of carbamate insecticides on biomass-derived compounds from lignin and furans for the first time. The synthesis pathway involves two steps in high yields. The novel compounds have also been purified and tested for insecticidal activity. The synthesis of these new compounds paves the way for a greener, safer alternative of insecticides.

536e Facile Conversion of Aqueous Phase Biomass Catalytic Pyrolysis Liquids into Hydrocarbons As a Wastewater Treatment Solution.

Foster Agblevor, Hossein Jahromi

The production of both conventional and catalytic pyrolysis oils always generates a liquid fraction that has very high water content and a very low organic fraction. This aqueous phase has to be treated for disposal in any large biofuels processing facility. Several methods have been proposed for disposing of such liquids. We investigated the hydrodeoxygenation (HDO) of aqueous-phase pinyon-juniper catalytic pyrolysis liquid using a new multi-functional red mud-supported nickel (Ni/RM) catalyst. The organic liquid yield after HDO of the liquid using 30 wt % Ni/RM at a reaction temperature of 350 °C was 47.8 wt % with an oxygen content of 1.14 wt %. The organic liquid fraction consisted of aliphatics, aromatics, and alkylated aromatic hydrocarbons which formed two phases with the aqueous phase. After decantation the aqueous phase contained more than 99% water. This method presents an alternative method for treating of the aqueous phase while simultaneously recovering the organic fraction as hydrocarbon fuels. The RM support catalyzed ketonization of carboxylic acids. The Ni metal catalyzed partial reduction of oxygenates that underwent carbonyl alkylation with aldehydes and ketones on the RM. Catalyst deactivation assessment suggested that oxidation and coke formation were the main controlling factors for deactivation of Ni and RM, respectively. For comparison, commercial (~65 wt %) $Ni/SiO_2-Al_2O_3$ was tested in HDO experiments, which gasified the soluble organics in the aqueous phase and did not produce liquid hydrocarbons but produced mostly methane.

536f Lignin Coupled with Catalyst Pyrolysis to Selective Production Chemicals: Influence of Lignin Structure and Properties of Catalyst.

Xiongjian DU, Shubin Wu

A new method for preparing phenols and aromatic hydrocarbons by in situ catalytic pyrolysis of pretreated lignin was studied. In order to further improve the selectivity of lignin pyrolysis products, due to the poor selectivity of current pyrolysis products, the influence of pyrolysis conditions on the formation of lignin pyrolysis products was investigated through the regulation of lignin structure, pyrolysis temperature and pyrolysis catalyst during pyrolysis. The Py-GC-MS analysis showed that phenol and its homologs accounted for 79.14% of the pyrolysis products of softwood kraft lignin catalyzed by TiO_2, and aromatic hydrocarbons accounted for 73.06% of the pyrolysis products of hydroxypropyl modified lignin catalyzed by HZSM. In addition, the pyrolysis of the lignin dimer model was also carried out to explore the pyrolysis reaction path of the lignin model, and the pyrolysis mechanism of the lignin model under the action of the catalyst was also described. These findings provide a new, green pathway for the production of high purity platform chemicals, which can ultimately facilitate the utilization of lignin through pyrolysis.

536g Development of a Method to Activate the By-Product (Biochar) of a Highly Porous Adsorption Substance Following Thermal Conversion of Biomass As a Renewable Energy Source

Ahmet Erdem, Al-Dahhan Muthanna, Zeyad Zeitoun

Research Interests

Biochar is a carbon-rich byproduct obtained from the degradation of lignocellulosic biomass as renewable energy sources to synthesis gas and/or liquid fuel by advanced thermal methods. The content of biochar generally contains carbon (%80-85) and other elements such as nitrogen, hydrogen, phosphorus, calcium, magnesium, and iron. It is possible to activate biochar by converting it into an efficient form and use it as a high porous adsorbent. The activation of biochar is achieved by increasing the surface area, number, and volume of micro-pores. Therefore, physical and/or chemical activation processes must be applied to obtain active biochar.

The current inability to utilize by-product biochar as an efficient product after the transformation of biomass with advanced thermal methods is a disadvantage in many respects, given the spread of such technologies. Although by-product biochar formed after the advanced thermal transformation of lignocellulosic biomass cannot be used as an adsorbent due to its lack of high surface area, a small amount can be used as a soil regulator and is usually consumed inefficiently in combustion boilers.

The dissertation titled "Production of Active Carbon from Biochar by Biomass Gasification" was examined as a doctoral study. In this study, the gasification of apricot seed and walnut shell biomass was carried out, and then a method and system were created by performing both physically and chemically the activation of the by-product biochar into activated carbon, which is a very valuable adsorbent and consumed a lot. However, studies in the literature have shown that biomass usually intended to be converted into activated carbon undergoes processes that create very high costs, making it uneconomical to apply these methods in industry on large scales.

Postdoctoral research aims to integrate the method created in the doctoral dissertation into advanced biomass thermal conversion processes and produce synthesis gas and/or liquid fuel from lignocellulosic biomass by intensifying the process. In addition, it aims to produce high-porous active biochar and investigate high-porous active biochar production methods in Missouri S&T laboratories.

In this study, the mass loss and surface properties of the produced activated biochar were examined. Subsequent analyses were carried out for materials that were deemed suitable for the use of activated carbon. It was analyzed with FT-IR Spectroscopy, XRF, EDS and SEM analyses were interpreted. Activated carbons, adsorption isotherms were examined depending on temperature, time and concentration. Adsorption equilibrium experimental data were investigated by examining their suitability for Langmuir and Freundlich models. Furthermore, kinetic studies on activated carbons were conducted, and calculations were made in certain models.

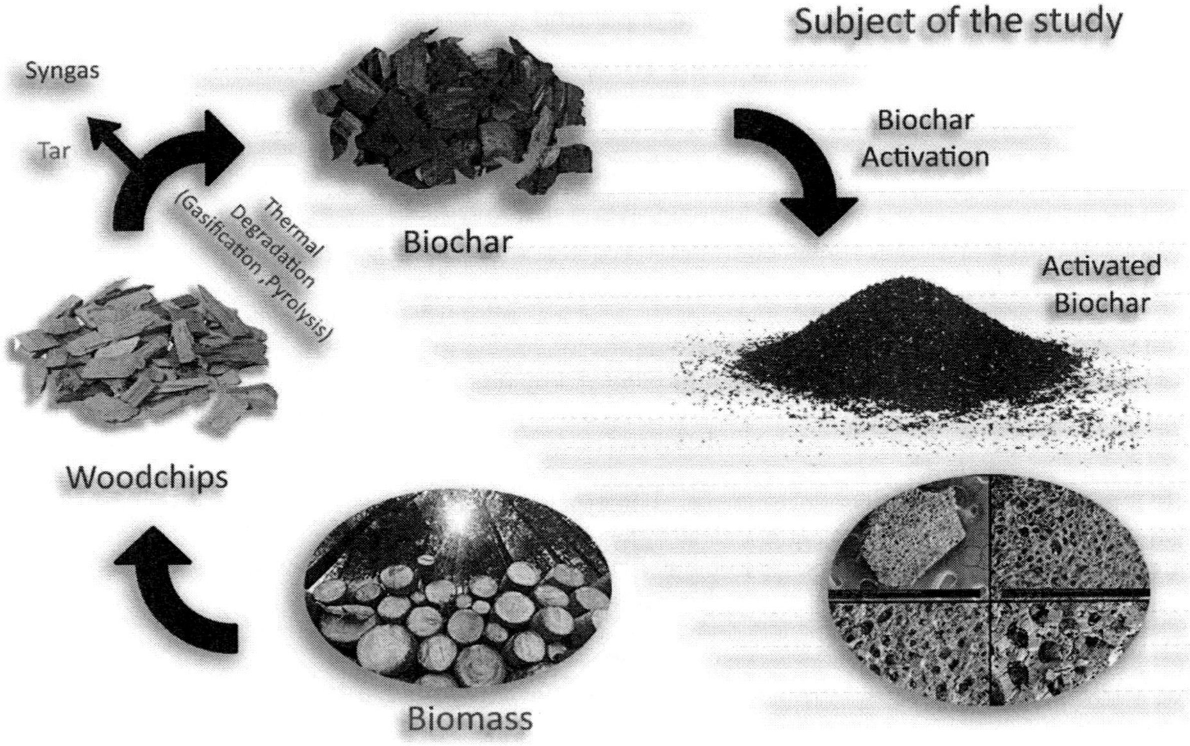

Subject of the study
Syngas
Tar
Thermal
Degradation
(Gasification ,Pyrolysis)
Biochar
Biochar
Activation
Activated
Biochar
Woodchips
Biomass

536h Removal of By-Product VOCs from Biomass Gasification Catalyzed By Ni/HZSM-5 : Optimization of Process Conditions

Liu Peng, Wenxuan Chen, Xueqin Li, Li Yanling, Yang Yantao, Sun Tanglei, Tingzhou Lei

Background, Aims and Scope. Biomass cracking and gasification technology, as a kind of thermochemical conversion technology, is widely used due to its advantages of strong raw material adaptability and high conversion efficiency. However, the organic pollutants produced in the biomass gasification process not only seriously affect the conversion efficiency, but also cause a certain degree of pollution to the atmospheric environment and endanger human health. These organic pollutants are condensable macromolecular gaseous organics produced during a series of complex gasification processes of biomass. Therefore, it can be considered that the organic pollutants produced in the biomass gasification process belong to the category of VOCs (Volatile Organic Compounds). At present, there are lots of VOCs treatment technologies, including adsorption, absorption, membrane separation, incineration, biodegradation, and catalysis. Many of these methods are prone to secondary pollution, poor stability and low removal efficiency during the treatment process, which are not conducive to the efficient elimination of VOCs. Catalytic method has high removal efficiency, does not cause secondary pollution, and is considered to be the most efficient removal method. For the VOCs produced by biomass gasification, catalytic cracking can be selected to remove them. In the method, a catalyst is added in the biomass gasification process, and the carbon-carbon bonds in the generated VOCs molecules are broken under the action of the catalyst, which reduces the activation energy of the cracking. This enables VOCs to be reformed and converted into small molecular gases at lower temperatures. This not only improves the removal efficiency of VOCs, but also further increases the yield of cracking gas. Undoubtedly, the core of catalytic cracking technology is the development of catalysts. Among various catalysts, nickel-based catalysts are the most widely used due to their low price and high activity. Active metal nickel has a certain activation effect on C-C bonds and C-H bonds in VOCs molecules, which can effectively promote the decomposition of organic macromolecules and effectively control the gas composition. The catalyst prepared with molecular sieve as carrier has large specific surface area, excellent thermal stability and special pore structure, so it has better catalytic activity for the catalytic cracking of VOCs. Previous studies have shown that HZSM-5 zeolite was considered as an ideal catalyst support due to its good surface acidity, hydrothermal stability and selectivity. Therefore, HZSM-5 molecular sieve is supported by active metal Ni, hoping to prepare a catalyst for the efficient removal of VOCs.

Methods. Molecular sieves were pretreated before catalyst preparation. After grinding the solid HZSM-5 molecular sieve into 80 mesh powder, it was calcined at 600 °C for 3 h, in order to improve the mechanical strength of the molecular sieve and remove impurities. The preparation of the nickel-based catalyst was based on HZSM-5 molecular sieve as the carrier, supported metal nickel as the active component, and traditional impregnation, ultrasonic assistance, ion exchange and rotary evaporation were used as auxiliary methods. The solids obtained by different auxiliary methods were dried in an oven at 105 °C for 12 h, ground into powder and calcined in a muffle furnace at 550 °C for 3 h to obtain the prepared Ni/HZSM-5 catalyst.

The performance test of the catalyst was carried out in a two-stage fixed bed reactor. The model compound was injected into the first stage reactor through a peristaltic pump, the catalyst was placed in the second stage, and the model compound was catalytically pyrolyzed under the action of the catalyst. In the whole process, high-purity N2 was used as the carrier gas, the gas product was collected by the air bag, and the liquid product was recovered after passing through the cooling system for subsequent qualitative and quantitative analysis.

Results and Discussion. In this study, the prepared nickel-based supported molecular sieve catalyst was used to study the catalytic cracking of VOCs model compounds. The influence of different catalyst preparation processes and cracking conditions was investigated, and the catalytic cracking conversion paths of various VOCs model compounds were explored. The conclusions are as follows: 1) The optimal preparation method of self-made nickel-based supported catalyst was ultrasonic-assisted excess impregnation method, and the optimal loading amount of Ni was 8 wt.%. The preparation method was convenient and easy to operate. Ultrasonic assistance not only promoted the refinement of the particle size of the active components and improved the dispersion of the active components, but also effectively controlled the loading of active metals through impregnation. 2) Through the study of cracking temperature, it was found that with the increase of temperature, the conversion rate of model substances and the yield of cracking gas increased significantly, and 800 °C was selected as the cracking temperature within the temperature selection range. 3) Catalytic cracking of various types of VOCs model compounds was carried out at 800 °C using Ni/HZMS-5 catalyst, and it was found that the model compounds could be converted from macromolecular organic compounds to small molecular gases under catalysis, and the conversion rate could mostly reach more than 90%. Model compounds such as toluene, phenol, and acetic acid can generate more hydrogen by cracking. The selectivity of H_2 and CH_4 from toluene cracking reached 93%, and cyclohexane reached more than 98%.

Conclusion. This study selected toluene and phenol as the representative of VOCs monocyclic aromatic hydrocarbon compounds, furan as the representative of heterocyclic compounds, and cyclohexane and acetic acid as the model compounds of aliphatic hydrocarbons. The catalytic cracking experiments were carried out using a self-made nickel-based supported molecular sieve catalyst to obtain the optimal preparation conditions of the catalyst. The transformation path of catalytic cracking of different VOCs model substances was explored, aiming to provide a new idea for the removal of VOCs and the efficient utilization of biomass resources.

536i Feasibility Assessment of Recycling Abandoned Aluminum Dross As Catalyst for Biomass Pyrolysis to Produce Hydrogen-Rich Gas

Xueqin Li, Liu Peng, Huang Sheng, Li Yanling, Yang Yantao, Wu Shiyong, Wu Youqing, Tingzhou Lei

Background, Aims and Scope. Aluminum dross is a hazardous solid waste that poses a severe challenge to the environment and public health. The conversion of biomass into hydrogen-rich gas is an effective way to address energy shortages and production. However, selecting suitable catalyst materials is the key to increasing the yield of hydrogen-rich gas, reducing pollutants such as tar, and determining whether hydrogen-rich gas can be successfully applied.

Methods. This study begins with the problems such as low activity, high cost, poor anti-coking properties, easy coking and poisoning, and short lifespan of existing catalysts for catalyzing the conversion of biomass into high-quality hydrogen-rich gas, followed by briefly introducing the characteristics, hazards, types, composition, and sources of aluminum dross. Next, with aluminum dross as a base material, we propose to preliminary synthesize carbon-aluminum composite materials by loading active components-carbon sources onto activated aluminum dross with a hydrothermal method. The hydrothermal method is utilized to load transition metal of iron onto the surface and pores of the selected carbon-aluminum composite material to prepare a carbon-aluminum composite material-supported iron catalyst. The pore structure, surface characteristics, and thermal stability of the catalyst are all characterized. Finally, using poplar sawdust as the raw material, the thermal decomposition product yield, gas release rate and yield are investigated as the main indicators to explore the impact of the catalyst on the thermal decomposition of biomass, and to comprehensively evaluate the feasibility of using aluminum dross to catalyze the conversion of biomass into hydrogen-rich gas.

Results and Discussion. The results indicate that the stability of activated aluminum dross is significantly improved, obtaining a larger specific surface area and pore volume. In particular, the acetic acid-activated aluminum dross has a specific surface area of 178.4 m^2/g, an average pore size of 9.40 nm, a pore volume of 0.42 cm^3/g, and a thermal conductivity coefficient of 0.3151 W/mK, which has similar pore structure characteristics to alumina oxide. In the carbon-aluminum composite material, the carbon sources are combined with the metallic oxides in the aluminum dross to form C-O-M clusters through hydrogen bonds or van der Waals forces, with a specific surface area, average pore size, and pore volume of 118 m^2/g, 10.3 nm, and 0.43 cm^3/g, respectively. It has a well-developed pore structure and a thermal conductivity coefficient of 0.2194 W/mK, and exhibits the advantages of both activated carbon and alumina oxide as a catalyst for catalytic biomass pyrolysis and gas reforming reaction, thus can be used as an excellent catalyst carrier for catalytic biomass pyrolysis to produce hydrogen-rich gas and gas reforming reaction. The carbon-aluminum composite material further loaded with iron exhibits a strong metal-support synergistic effect on the surface. This enhances the ability of the aluminum dross to promote reforming and cracking reactions, leading to the cleavage of heavy chain volatiles into lighter molecules and gases, resulting in a higher gas yield (43.77%) and lower coke deposition (19.79%). The H_2 yield is significantly improved to 64.24 $ml/g_{biomass}$, providing a promising route for the pyrolysis of biomass to produce hydrogen-rich gas.

Conclusion. It is feasible to recycle abandoned aluminum dross as catalyst for biomass pyrolysis to produce hydrogen-rich gas. This approach represents the reuse of waste resources and can effectively promote the efficient conversion of biomass and by-products, laying a theoretical foundation for the application of waste resources in the fields of environment, energy, and chemical industry, achieving the dual goals of protecting natural resources and obtaining economic benefits.

AUTHOR INDEX

AIChE
120 Wall Street, FL 23
New York, NY 10005-4020

ISBN 978-1-7138-9295-3